Chemical Elements

2nd Edition

Chemical Elements

2nd Edition VOLUME 1: A-F

David E. Newton

Kathleen J. Edgar, Editor

U·X·L
A part of Gale, Cengage Learning

GALE
CENGAGE Learning™

Detroit • New York • San Francisco • New Haven, Conn • Waterville, Maine • London

Chemical Elements, 2nd Edition

David E. Newton

Project Editor: Kathleen J. Edgar

Managing Editor: Debra Kirby

Rights Acquisition and Management: Barbara McNeil, Jackie Jones, and Robyn Young

Imaging and Multimedia: John L. Watkins

Composition: Evi Abou-El-Seoud, Mary Beth Trimper

Manufacturing: Wendy Blurton, Dorothy Maki

Product Managers: Julia Furtaw, John Boyless

Product Design: Jennifer Wahi

For product information and technology assistance, contact us at **Gale Customer Support, 1-800-877-4253.**
For permission to use material from this text or product, submit all requests online at **www.cengage.com/permissions.**
Further permissions questions can be e-mailed to **permissionrequest@cengage.com.**

Cover photographs: Beaker image by originalpunkt, used under license from Shutterstock.com. All other photos by Photos.com/Getty Images.

LIBRARY OF CONGRESS CATALOGING-IN-PUBLICATION DATA

Newton, David E.
  Chemical elements / David E. Newton ; Kathleen J. Edgar, editor. -- 2nd ed.
    p. cm.
  Includes bibliographical references and index.
  ISBN-13: 978-1-4144-7608-7 (set)
  ISBN-13: 978-1-4144-7609-4 (v. 1)
  ISBN-13: 978-1-4144-7610-0 (v. 2)
  ISBN-13: 978-1-4144-7611-7 (v. 3)
  [etc.]
  1. Chemical elements.   I. Edgar, Kathleen J.   II. Title.
  QD466.N464 2010
  546--dc22   2009050931

Gale
27500 Drake Rd.
Farmington Hills, MI, 48331-3535

ISBN-13: 978-1-4144-7608-7 (set)   ISBN-10: 1-4144-7608-6 (set)
ISBN-13: 978-1-4144-7609-4 (vol. 1)   ISBN-10: 1-4144-7609-4 (vol. 1)
ISBN-13: 978-1-4144-7610-0 (vol. 2)   ISBN-10: 1-4144-7610-8 (vol. 2)
ISBN-13: 978-1-4144-7611-7 (vol. 3)   ISBN-10: 1-4144-7611-6 (vol. 3)

This title is also available as an e-book.
ISBN-13: 978-1-4144-7612-4 (set)   ISBN-10: 1-4144-7612-4 (set)
Contact your Gale sales representative for ordering information.

Printed by China Translation & Printing Services Limited,
Guangdong Province, China. 2nd printing. 07/2011
2 3 4 5 6 7 14 13 12 11

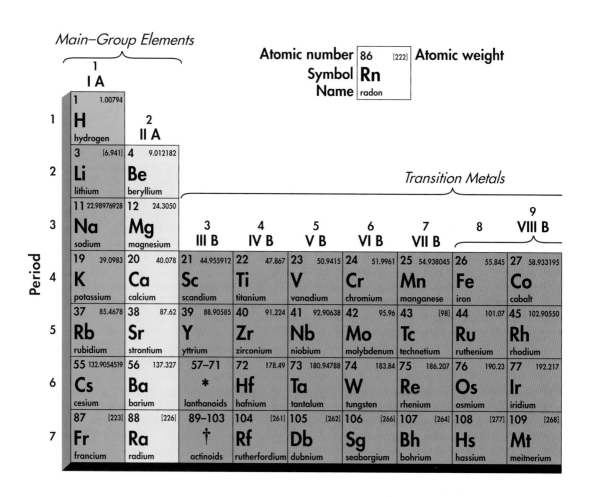

Main–Group Elements

Atomic number | 86 | [222] | Atomic weight
Symbol | Rn
Name | radon

Transition Metals

	1 IA	2 IIA	3 IIIB	4 IVB	5 VB	6 VIB	7 VIIB	8	9 VIIIB
1	1 1.00794 H hydrogen								
2	3 [6.941] Li lithium	4 9.012182 Be beryllium							
3	11 22.98976928 Na sodium	12 24.3050 Mg magnesium							
4	19 39.0983 K potassium	20 40.078 Ca calcium	21 44.955912 Sc scandium	22 47.867 Ti titanium	23 50.9415 V vanadium	24 51.9961 Cr chromium	25 54.938045 Mn manganese	26 55.845 Fe iron	27 58.933195 Co cobalt
5	37 85.4678 Rb rubidium	38 87.62 Sr strontium	39 88.90585 Y yttrium	40 91.224 Zr zirconium	41 92.90638 Nb niobium	42 95.96 Mo molybdenum	43 [98] Tc technetium	44 101.07 Ru ruthenium	45 102.90550 Rh rhodium
6	55 132.9054519 Cs cesium	56 137.327 Ba barium	57–71 * lanthanoids	72 178.49 Hf hafnium	73 180.94788 Ta tantalum	74 183.84 W tungsten	75 186.207 Re rhenium	76 190.23 Os osmium	77 192.217 Ir iridium
7	87 [223] Fr francium	88 [226] Ra radium	89–103 † actinoids	104 [261] Rf rutherfordium	105 [262] Db dubnium	106 [266] Sg seaborgium	107 [264] Bh bohrium	108 [277] Hs hassium	109 [268] Mt meitnerium

Period

*Lanthanoids

57 138.90547 La lanthanum	58 140.116 Ce cerium	59 140.90765 Pr praseodymium	60 144.242 Nd neodymium	61 [145] Pm promethium	62 150.36 Sm samarium
89 [227] Ac actinium	90 232.03806 Th thorium	91 231.03588 Pa protactinium	92 238.02891 U uranium	93 [237] Np neptunium	94 [244] Pu plutonium

†Actinoids

Color Key

Main–Group Elements

13 III A	14 IV A	15 V A	16 VI A	17 VII A	18 VIII A
					2 4.002602 **He** helium
5 10.811 **B** boron	6 12.0107 **C** carbon	7 14.0067 **N** nitrogen	8 15.9994 **O** oxygen	9 18.9984032 **F** fluorine	10 20.1797 **Ne** neon
13 26.9815386 **Al** aluminum	14 28.0855 **Si** silicon	15 30.973762 **P** phosphorus	16 32.065 **S** sulfur	17 35.453 **Cl** chlorine	18 39.948 **Ar** argon

10	11 I B	12 II B	13 III A	14 IV A	15 V A	16 VI A	17 VII A	18 VIII A
28 58.6934 **Ni** nickel	29 63.546 **Cu** copper	30 65.38 **Zn** zinc	31 69.723 **Ga** gallium	32 72.64 **Ge** germanium	33 74.92160 **As** arsenic	34 78.96 **Se** selenium	35 79.904 **Br** bromine	36 83.798 **Kr** krypton
46 106.42 **Pd** palladium	47 107.8682 **Ag** silver	48 112.411 **Cd** cadmium	49 114.818 **In** indium	50 118.710 **Sn** tin	51 121.760 **Sb** antimony	52 127.60 **Te** tellurium	53 126.90447 **I** iodine	54 131.293 **Xe** xenon
78 195.084 **Pt** platinum	79 196.966569 **Au** gold	80 200.59 **Hg** mercury	81 204.3833 **Tl** thallium	82 207.2 **Pb** lead	83 208.98040 **Bi** bismuth	84 [209] **Po** polonium	85 [210] **At** astatine	86 [222] **Rn** radon
110 [271] **Ds** darmstadtium	111 [272] **Rg** roentgenium	112 [285] **Cn** copernicium	113 [284] **Uut** ununtrium	114 [289] **Uuq** ununquadium	115 [288] **Uup** ununpentium	116 [293] **Uuh** ununhexium	117 **Uus** ununseptium	118 [294] **Uuo** ununoctium

Inner–Transition Metals

63 151.964 **Eu** europium	64 157.25 **Gd** gadolinium	65 158.92535 **Tb** terbium	66 162.500 **Dy** dysprosium	67 164.93032 **Ho** holmium	68 167.259 **Er** erbium	69 168.93421 **Tm** thulium	70 173.054 **Yb** ytterbium	71 174.9668 **Lu** lutetium
95 [243] **Am** americium	96 [247] **Cm** curium	97 [247] **Bk** berkelium	98 [251] **Cf** californium	99 [252] **Es** einsteinium	100 [257] **Fm** fermium	101 [258] **Md** mendelevium	102 [259] **No** nobelium	103 [262] **Lr** lawrencium

Contents

VOLUME 2

Contents: Elements by Atomic Number

Elements are listed in order by atomic number. Volume numbers appear in **_bold-italic type_** followed by the page numbers.

Chemical Elements, 2nd Edition

Contents: Elements by Family Group

Bold-Italic type indicates
volume numbers.

Chemical Elements, 2nd Edition

LANTHANOIDS

ACTINOIDS

Reader's Guide

Many young people like to play with Lego blocks, erector sets, and similar building games, including virtual games found on the Internet. It's fun to see how many different ways a few simple shapes can be put together.

The same can be said of chemistry. The world is filled with an untold number of different objects, ranging from crystals and snowflakes to plant and animal cells to plastics and medicines. Yet all of those objects are made from various combinations of only about 100 basic materials: the chemical elements.

Scientists have been intrigued about the idea of an "element" for more than 2,000 years. The early Greeks developed complicated schemes that explained everything in nature using only a few basic materials, such as earth, air, fire, and water. The Greeks were wrong in terms of the materials they believed to be "elemental." But they were on the right track in developing the concept that such materials did exist.

By the 1600s, chemists were just beginning to develop a modern definition of an element. An element, they said, was any object that cannot be reduced to some simpler form of matter. Over the next 300 years, research showed that about 100 such materials did exist. These materials range from such well-known elements as oxygen, hydrogen, iron, gold, and silver to substances that are not at all well known, elements such as neodymium, terbium, rhenium, seaborgium, darmstadtium, and copernicium.

By the mid-1800s, the search for new chemical elements had created a new problem. About 50 elements were known at the time. But no one yet knew how these different elements related to each other, if they did at all. Then, in one of the great coincidences in chemical history, that question was answered independently by two scientists at almost the same time, German chemist Lothar Meyer (1830–1895) and Russian chemist Dmitri Mendeleev (1834–1907). (Meyer, however, did not publish his research until 1870, nor did he predict the existence of undiscovered elements as Mendeleev did.)

Meyer and Mendeleev discovered that the elements could be grouped together to make them easier to study. The grouping occurred naturally when the elements were laid out in order, according to their atomic weight. Atomic weight is a quantity indicating atomic mass that tells how much matter there is in an element or how dense it is. The product of Meyer and Mendeleev's research is one of the most famous visual aids in all of science, the periodic table. Nearly every classroom has a copy of this table. (A copy is available in this book just before the table of contents.) It lists all of the known chemical elements, arranged in rows and columns. The elements that lie within a single column or a single row all have characteristics that relate to each other. Chemists and students of chemistry use the periodic table to better understand individual elements and the way the elements are similar to and different from each other.

About *Chemical Elements, 2nd Edition*

Chemical Elements, 2nd Edition is designed as an introduction to the chemical elements. This new edition, presented in full color, updates the earlier three-volume set, providing new information about the elements as well as many new photographs.

Students will find *Chemical Elements* useful in a number of ways. First, it is a valuable source of fundamental information for research reports, science fair projects, classroom demonstrations, and other activities. Second, it provides more detail about elements and compounds that are only mentioned in other science textbooks or classrooms. Third, it is an interesting source of information about the building blocks of nature for those who simply want to know more about the elements.

Elements with atomic numbers 1 through 100 are presented in separate entries. The transfermium elements (elements 101 through 112) are covered in one entry, which also discusses six additional elements (113, 114, 115, 116, 118, and 122) that have yet to be confirmed by the International Union of Pure and Applied Chemistry (IUPAC).

Entry Format

Chemical Elements is arranged alphabetically by element name. All entries open with an overview section designed to introduce students to the basics of the element. Each entry then contains specific information in the following categories:

- Discovery and Naming
- Physical Properties
- Chemical Properties
- Occurrence in Nature
- Isotopes
- Extraction
- Uses
- Compounds
- Health Effects

In addition, the first page of each entry features a Key Facts section in the margin. Here, the element's chemical symbol, atomic number, atomic mass, family, and pronunciation are shown. A diagram of an atom of the element is also shown at the top of the page. In the diagram, the atom's electrons are arranged in various "energy levels" outside the nucleus. Inside the nucleus, the number of protons and neutrons is indicated.

Entries are easy to read and written in a straightforward style. Difficult words are defined within the text. Each entry also includes a "Words to Know" sidebar that defines technical words and scientific terms. This enables students to learn vocabulary appropriate to chemistry without having to consult other sources for definitions.

Special Features

Chemical Elements, 2nd Edition includes a number of special features that help make the connection between the elements, minerals, people who discovered and worked with them, and common uses of the elements. These features include:

- Nearly 300 photographs and illustrations, most in color, showcasing various elements, the products in which they are used, and the places in which they are found. Such imagery brings the elements to life. Black-and-white images of historical figures are also included.

- Sidebars discussing fascinating supplemental information about scientists, theories, uses of elements, and more.
- Extensive cross-references. Other elements mentioned within an entry appear in bold type upon first mention, serving as a helpful reminder that separate entries are written about these other elements.
- A periodic table (positioned right before the table of contents) that includes the following for each element: name, symbol, atomic number, and atomic mass. A color key also informs students about the various groupings of the elements.
- Three tables of contents—alphabetically by element name, by atomic number, and by family group. This presentation provides varied access to the elements.
- A timeline in each volume. This section details the chronology of the discovery of the elements.
- A cumulative Words to Know section in each volume. This section collects all the glossary terms used in the individual entries.
- A Where to Learn More section at the back of each volume. This bibliography provides information on books, periodicals, Web sites, and organizations that may be useful to those interested in learning more about the chemical elements.
- A comprehensive index that quickly points readers to the elements, minerals, and people mentioned in *Chemical Elements, 2nd Edition.*

A Note about Isotopes and Atomic Weights

Various authorities list slightly different isotopes for some elements. One reason is that the discovery of a new isotope may not yet have been confirmed by other researchers, so its existence is uncertain. Lists of isotopes may change also because new isotopes are being discovered from time to time. The number of isotopes, with a known half life, listed in this book is based on information available from the Lawrence Berkeley National Laboratory in Berkeley, California. The list is up to date as of December 2009. All stable isotopes exist in nature. Some radioactive isotopes also exist in nature, but the vast majority has been prepared synthetically in the laboratory.

The atomic weight information presented in the entries and periodic table is the latest available (at press time) from the International Union of

Pure and Applied Chemistry (IUPAC), the governing body that officially confirms and names any new elements. The IUPAC lists the atomic weight in brackets when an element does not have any stable nuclides. For example, the atomic weight of actinium is listed as [227]. This represents the mass of its longest-lived isotope.

Special Thanks

The project editor wishes to thank graphic artist Christine O'Bryan for the amazing job she did colorizing the atom diagrams and updating the periodic table. Thanks also to Robyn Young for working her magic in acquiring new images for this edition. Also thanks to Lemma Shomali for lending her content expertise to this set. All your efforts are greatly appreciated.

Comments and Suggestions

We welcome your comments on this work as well as suggestions for future science titles. Please write: Editors, *Chemical Elements, 2nd Edition*, U•X•L, 27500 Drake Rd., Farmington Hills, MI, 48331-3535; call toll-free: 800-347-4253; send fax to 248-699-8066; or send e-mail via http://www.gale.com.

Timeline: The Discovery of Elements

Assigning credit for the discovery of a new element is often a difficult and complicated process. First, many elements were in use well before recorded history. In some cases, these elements were known in the form of their compounds, but not as pure elements. Elements that fall into this category include carbon, copper, gold, iron, lead, mercury, silver, sulfur, tin, and, perhaps, zinc.

In addition, the discovery of an element has seldom been a single, clear-cut event that occurs in such a way that everyone agrees that "X" should receive credit for discovering the element. Instead, the first step in the process of discovery is often the recognition that a new substance has been found—a new mineral, rock, compound, or other material—that has properties different from anything previously known. This discovery may lead a chemist (or a number of chemists) to suspect the existence of a new element.

The next step may be to isolate the element, either in its pure form or, more commonly, as a compound, such as the oxide or sulfide of the new element. Finally, someone is able to prepare a pure sample of the element, which the world then sees for the first time. An example of this sequence of events can be seen in the elements that make up groups 1 and 2 of the periodic table. Most of those elements were known in one form or another for centuries. But it was not until the early 1800s that Sir Humphry Davy found a method for isolating the pure elements from their oxides.

The process becomes even more complicated when a truly new element is discovered which, sometime later, is found not to be a single element, but a mixture of two or more new elements. The story of the discovery of the rare earth elements is probably the best example of this process.

This sequence of events often takes place over an extended period of time, many years or even decades. For that reason, assigning a specific date to the discovery of an element can also be difficult. Does one choose the date and person when a new compound of the element is discovered, when the pure element itself is prepared, when the discoverer publicly announces his or her discovery, or when official confirmation of the discovery is announced?

For all these reasons, the dates and names listed below must be considered as somewhat ambiguous. For more detailed information about the discovery of each element, the reader should refer to the entry for that element in the main body of this set of books.

About 800 CE Persian natural philosopher Abu Musa Jābir ibn Hayyān al azdi (better known as Geber) is credited with discovering **antimony, arsenic,** and **bismuth**.

About 800 CE Indian metallurgist Rasaratna Samuchaya is perhaps the first person to recognize **zinc** as an element.

1250 German natural philosopher Albertus Magnus is credited as being the first European to discover **arsenic**.

About 1450 The apocryphal Basilius Valentinus (Basil Valentine) is the first European to mention elemental **antimony** and **bismuth**.

1526 Swiss physician Auroleus Phillipus Theostratus Bombastus von Hohenheim (Paracelsus) is acknowledged as the modern discoverer of **zinc**.

1669 German physician Hennig Brand discovers **phosphorus**.

1735 Swedish chemist Georg Brandt discovers **cobalt**.

1735–1748 Spanish military Leader Don Antonio de Ulloa discovers **platinum**.

1751 Swedish mineralogist Axel Fredrik Cronstedt discovers **nickel**.

1755 Scottish physician and chemist Joseph Black recognizes the presence of a new element in magnesia alba, later found to be **magnesium**.

1766 English chemist and physicist Henry Cavendish discovers **hydrogen**.

1771 Swedish chemist Carl Wilhelm Scheele discovers **oxygen**, but does not publish his discovery until 1777.

1772 Scottish physician and chemist Daniel Rutherford discovers **nitrogen**.

1774 Swedish chemist Carl Wilhelm Scheele discovers **chlorine**.

1774 Swedish mineralogist Johann Gottlieb Gahn discovers **manganese**.

1774 English chemist Joseph Priestley discovers **oxygen** and, because he announces his results almost immediately, is often credited as the discoverer of the element.

1781 Swedish chemist Peter Jacob Hjelm discovers **molybdenum**.

1782 Austrian mineralogist Baron Franz Joseph Müller von Reichenstein discovers **tellurium**.

1783 Spanish scientists Don Fausto D'Elhuyard and Don Juan José D'Elhuyard and Swedish chemist Carl Wilhelm Scheele discover **tungsten**.

1787 Scottish military surgeon William Cruikshank and Irish chemist and physicist Adair Crawford independently announce the probable existence of a new element, later found to be **strontium**.

1789 German chemist Martin Klaproth recognizes the presence of **uranium** in pitchblende, but does not isolate the element.

1789 German chemist Martin Klaproth discovers **zirconium**. The element is not isolated until 1824.

1791 English clergyman William Gregor discovers an oxide of **titanium**. German chemist Martin Klaproth makes a similar discovery four years later. The element is not isolated until 1910.

1794 Finnish chemist Johan Gadolin discovers **yttrium**.

1797 French chemist Louis-Nicolas Vauquelin discovers **chromium**.

1798 French chemist Louis-Nicolas Vauquelin discovers **beryllium**.

1801 English chemist Charles Hatchett discovers **niobium**.

1801 Spanish-Mexican metallurgist Andrés Manuel del Río discovers **vanadium**.

1802 Swedish chemist and mineralogist Anders Gustaf Ekeberg discovers **tantalum**.

1803 English chemist and physicist William Hyde Wollaston discovers **palladium**.

1803 Swedish chemists Jöns Jakob Berzelius and Wilhelm Hisinger and German chemist Martin Klaproth discover the black rock of Bastnas, Sweden, which led to the discovery of several elements. Berzelius and Hisinger originally assume the rock is a new element, which they name **cerium**.

1803 English chemist Smithson Tennant discovers **osmium** and **iridium**.

1804 English chemist and physicist William Hyde Wollaston discovers **rhodium**.

1807–1808 English chemist Sir Humphry Davy isolates a number of elements in a pure form for the first time, including **potassium, sodium, magnesium, barium, calcium**, and **strontium**.

1808 French chemists Louis Jacques Thênard and Joseph Louis Gay-Lussac discover **boron**. Davy isolates the element a few days after its discovery has been announced.

1811 French chemist Bernard Courtois discovers **iodine**.

1817 Swedish chemist Johan August Arfwedson discovers **lithium**.

1817 German chemists Friedrich Stromeyer, Karl Samuel Leberecht Hermann, and J. C. H. Roloff independently discover **cadmium**, a name chosen by Stromeyer.

1818 Swedish chemists Jöns Jakob Berzelius and J. G. Gahn discover **selenium**.

1823 Swedish chemist Jöns Jakob Berzelius discovers **silicon**.

1825 Danish chemist and physicist Hans Christian Oersted discovers **aluminum**.

1825 French chemist Antoine-Jérôme Balard and German chemist Leopold Gmelin independently discover **bromine**.

1829 Swedish chemist Jöns Jakob Berzelius discovers **thorium**.

1830 Swedish chemist Nils Gabriel Sefström rediscovers **vanadium**.

1838 Swedish chemist Carl Gustav Mosander discovers that **cerium** contains a new element, which he names **lanthanum**. His lanthanum is later found to consist of four new elements.

1842 Swedish chemist Carl Gustav Mosander discovers that the earth called yttria actually consists of two new elements, **erbium** and **terbium**.

1844 Russian chemist Carl Ernst Claus discovers **ruthenium**.

1860 German chemists Robert Bunsen and Gustav Kirchhoff discover **cesium**.

1861 German chemists Robert Bunsen and Gustav Kirchhoff discover **rubidium**.

1861 British physicist Sir William Crookes discovers **thallium**.

1863 German chemists Ferdinand Reich and Hieronymus Theodor Richter discover **indium**.

1868 Pierre Janssen and Norman Lockyer discover **helium** in the spectrum of the sun.

1875 Paul-Émile Lecoq de Boisbaudran discovers **gallium**.

1878 Swiss chemist Jean-Charles-Galissard de Marignac, Swedish chemist Lars Fredrik Nilson, and French chemist Georges Urbain all receive partial credit for the discovery of **ytterbium**.

1878–1879 Swedish chemist Per Teodor Cleve discovers **holmium** and **thulium**.

1879 Swedish chemist Lars Fredrik Nilson discovers **scandium**.

1880 French chemist Paul-Émile Lecoq de Boisbaudran discovers **samarium**.

1880 French chemist Jean-Charles-Galissard de Marignac discovers **gadolinium**.

1885 Austrian chemist Carl Auer von Welsbach discovers **praseodymium** and **neodymium**.

1885 German chemist Clemens Alexander Winkler discovers **germanium**.

1886 French chemist Henri Moissan discovers **fluorine**.

1886 French chemist Paul-Émile Lecoq de Boisbaudran discovers **dysprosium**.

1894 English chemists Lord Rayleigh and Sir William Ramsay discover **argon**.

1895 English chemist Sir William Ramsay and Swedish chemists Per Teodor Cleve and Nils Abraham Langlet independently discover **helium** in the mineral clevite, the first discovery of the element on Earth's surface.

1896 French chemist Eugène-Anatole Demarçay discovers **europium**.

1898 English chemists Sir William Ramsay and Morris Travers discover **krypton, neon**, and **xenon**.

1898 French physicists Marie and Pierre Curie discover **polonium** and **radium**.

1898 German physicist Friedrich Ernst Dorn discovers **radon**.

1899 French chemist André Debierne discovers **actinium**.

1906 French chemist Georges Urbain and Austrian chemist Carl Auer von Welsbach independently discover **lutetium**.

1908 Japanese chemist Masataka Ogawa discovers **rhenium**, but erroneously assigns it to atomic number 43, instead of its correct atomic number of 75. He names the element nipponium, but his research is forgotten and ignored for many years.

1911 French chemist Georges Urbain and Russian chemist Vladimir Ivanovich Vernadskij independently discover **hafnium**. Their research is unconfirmed because of World War I (1914–1918).

1917 Three research teams, consisting of German physicists Lise Meitner and Otto Hahn, Polish-American physical chemist Kasimir Fajans and German chemist O. H. Göhring, and English physicists Frederick Soddy and John A. Cranston, independently and almost simultaneously discover **protactinium**.

1923 Dutch physicist Dirk Coster and Hungarian chemist George Charles de Hevesy rediscover **hafnium** and are generally recognized as discoverers of the element.

1925 German chemists Walter Noddack, Ida Tacke, and Otto Berg rediscover and name **rhenium**.

1937 Italian physicist Emilio Segré and his colleague Carlo Perrier discover **technetium**.

1939 French chemist Marguerite Perey discovers **francium**.

1940 Dale R. Corson, Kenneth R. Mackenzie, and Emilio Segré discover **astatine**.

1940 Edwin M. McMillan and Philip H. Abelson prepare **neptunium**.

1940 University of California at Berkeley researchers Glenn Seaborg, Arthur C. Wahl, J. K. Kennedy, and E. M. McMillan prepare **plutonium**.

1944 University of California at Berkeley researchers Glenn Seaborg, Albert Ghiorso, Ralph A. James, and Leon O. Morgan prepare **americium**.

1944 University of California at Berkeley researchers Glenn Seaborg, Albert Ghiorso, and Ralph A. James prepare **curium**.

1945 Oak Ridge National Laboratory researchers Jacob A. Marinsky, Lawrence E. Glendenin, and Charles D. Coryell discover **promethium**. The element had probably been found as early as 1942 by Chien Shiung Wu, Emilio Segré, and Hans Bethe.

1949 University of California at Berkeley researchers Stanley G. Thompson, Albert Ghiorso, and Glenn Seaborg prepare **berkelium**.

1950 University of California at Berkeley researchers Glenn Seaborg, Albert Ghiorso, Kenneth Street Jr., and Stanley G. Thompson prepare **californium**.

1952 A team of University of California at Berkeley researchers led by Albert Ghiorso prepares **einsteinium** (#99) and **fermium** (#100).

1955 University of California at Berkeley researchers Glenn Seaborg, Albert Ghiorso, Bernard G. Harvey, Gregory R. Choppin, and Stanley G. Thompson prepare **mendelevium** (#101).

1958 University of California at Berkeley researchers Glenn Seaborg, Albert Ghiorso, Tørbjorn Sikkeland, and J. R. Walton prepare **nobelium** (#102).

1961 University of California at Berkeley researchers Albert Ghiorso, Tørbjorn Sikkeland, Almon E. Larsh, and R. M. Latimer prepare **lawrencium** (#103).

1964 A research team at the Joint Institute for Nuclear Research at Dubna, Russia, led by Russian physicist Georgy Nikolaevich Flerov, produces **rutherfordium** (#104).

1968 A research team at the Joint Institute for Nuclear Research at Dubna, Russia, led by Russian physicist Georgy Nikolaevich Flerov, produces **dubnium** (#105).

1974 A research team at the University of California at Berkeley led by Albert Ghiorso produces **seaborgium** (#106).

1981 A research team led by Peter Armbruster and Gottfried Münzenberg at the Gesellschaft für Schwerionenforschung (Institute for Heavy Ion Research) in Darmstadt, Germany, produces **bohrium** (#107).

1982 The Heavy Ion Research team in Darmstadt produces **meitnerium** (#109).

1984 The Heavy Ion Research team in Darmstadt produces **hassium** (#108).

1994 The Heavy Ion Research team in Darmstadt, under the leadership of Sigurd Hofmann, produces **darmstadtium** (#110) and **roentgenium** (#111).

1996 The Heavy Ion Research team in Darmstadt, under the leadership of Sigurd Hofmann, produces **copernicium** (#112).

1999 Researchers at the Joint Institute for Nuclear Research announce the discovery of element #114, ununquadium.

2000 Researchers at the Joint Institute for Nuclear Research announce the discovery of element #116, ununhexium. This and all subsequent discoveries have not been confirmed nor a name approved for the possible new elements as of early 2010.

2002 Researchers at the Joint Institute for Nuclear Research at Dubna and the Lawrence Livermore National Laboratory in California announce the discovery of element #118, ununoctium.

2003 Researchers at the Joint Institute for Nuclear Research at Dubna and the Lawrence Livermore National Laboratory in California announce the discovery of element #113 ununtrium and element #115 ununpentium.

2008 Researchers at the Hebrew University at Jerusalem, under the leadership of Amnon Marinov, report the discovery of single atoms of element #122, unbibium, in naturally occurring deposits of **thorium**. Although unconfirmed as of early 2010, this report would be the first discovery of a new element in nature since 1939 (**francium**).

2009 The International Union of Pure and Applied Chemistry (IUPAC) officially recognizes element 112, Copernicium (Cn), and adds the name to the standard periodic table.

2009 A team of researchers at the Lawrence Berkeley National Laboratory confirm the production of #114, ununquadium.

Words to Know

Abrasive: A powdery material used to grind or polish other materials.

Absolute zero: The lowest temperature possible, about -459°F (-273°C).

Actinoid family: Formerly Actinide family; elements in the periodic table with atomic numbers 90 through 103.

Alchemy: A kind of pre-science that existed from about 500 BCE to about the end of the 16th century.

Alkali: A chemical with properties opposite those of an acid.

Alkali metal: An element in Group 1 (IA) of the periodic table.

Alkaline earth metal: An element found in Group 2 (IIA) of the periodic table.

Allotropes: Forms of an element with different physical and chemical properties.

Alloy: A mixture of two or more metals that has properties different from those of the individual metals.

Alpha particles: Tiny, atom-sized particles that can destroy cells.

Alpha radiation: A form of radiation that consists of very fast moving alpha particles and helium atoms without their electrons.

Amalgam: A combination of mercury and at least one other metal.

Amorphous: Without crystalline shape.

Anhydrous ammonia: Dry ammonia gas.

Antiseptic: A chemical that stops the growth of germs.

Aqua regia: A mixture of hydrochloric and nitric acids that often reacts with materials that do not react with either acid separately.

Battery: A device for changing chemical energy into electrical energy.

Biochemistry: The field of chemistry concerned with the study of compounds found in living organisms.

Biocompatible: Not causing a reaction when placed into the body.

Bipolar disorder: A condition in which a person experiences wild mood swings.

Brass: An alloy of copper and zinc.

Bronze: An alloy of copper and tin.

Bronze Age: A period in human history ranging from about 3500 to 1000 BCE, when bronze was widely used for weapons, utensils, and ornamental objects.

Buckminsterfullerene: Full name for buckyball or fullerene; also see Buckyball.

Buckyball: An allotrope of carbon whose 60 carbon atoms are arranged in a sphere-like form.

Capacitor: An electrical device, somewhat like a battery, that collects and then stores up electrical charges.

Carat: A unit of weight for gold and other precious metals, equal to one fifth of a gram, or 200 milligrams.

Carbon arc lamp: A lamp for producing very bright white light.

Carbon-14 dating: A technique that allows archaeologists to estimate the age of once-living materials by using the knowledge that carbon-14 is found in all living carbon materials; once an organism dies, no more carbon-14 remains.

Cassiterite: An ore of tin containing tin oxide, the major commercial source of tin metal.

Catalyst: A substance used to speed up a chemical reaction without undergoing any change itself.

Chalcogens: Elements in Group 16 (VIA) of the periodic table.

Chemical reagent: A substance, such as an acid or an alkali, used to study other substances.

Chlorofluorocarbons (CFCs): A family of chemical compounds consisting of carbon, fluorine, and chlorine that were once used widely as propellants in commercial sprays but regulated in the United States since 1987 because of their harmful environmental effects.

Corrosive agent: A material that tends to vigorously react or eat away at something.

Cyclotron: A particle accelerator, or "atom smasher," in which small particles, such as protons, are made to travel very fast and then collide with atoms, causing the atoms to break apart.

Density: The mass of a substance per unit volume.

Diagnosis: Finding out what medical problems a person may have.

Distillation: A process by which two or more liquids can be separated from each other by heating them to their boiling points.

"Doped": Containing a small amount of a material as an impurity.

Ductile: Capable of being drawn into thin wires.

Earth: In mineralogy, a naturally occurring form of an element, often an oxide of the element.

Electrolysis: A process by which a compound is broken down by passing an electric current through it.

Electroplating: The process by which a thin layer of one metal is laid down on top of a second metal.

Enzyme: A substance that stimulates certain chemical reactions in the body.

Fabrication: Shaping, molding, bending, cutting, and working with a metal.

Fission: The process by which large atoms break apart, releasing large amounts of energy, smaller atoms, and neutrons in the process.

Fly ash: The powdery material produced during the production of iron or some other metal.

Frasch method: A method for removing sulfur from underground mines by pumping hot air and water down a set of pipes.

Fuel cell: Any system that uses chemical reactions to produce electricity.

Fullerene: Alternative name for buckyball; also see Buckyball.

Galvanizing: The process of laying down a thin layer of zinc on the surface of a second metal.

Gamma rays: A form of radiation similar to X rays.

Global warming: A phenomenon in which the average temperature of Earth rises, melting icecaps, raising sea levels, and causing other environmental problems. Causes include human-activities, including heavy emissions of carbon dioxide (CO_2).

Half life: The time it takes for half of a sample of a radioactive element to break down.

Halogen: One of the elements in Group 17 (VIIA) of the periodic table.

Heat exchange medium: A material that picks up heat in one place and carries it to another place.

Hydrocarbons: Compounds made of carbon and hydrogen.

Hypoallergenic: Not causing an allergic reaction.

Inactive: Does not react with any other element.

Inert: Not very active.

Isotope: Two or more forms of an element that differ from each other according to their mass number.

Lanthanoid family: Formerly Lanthanide family; the elements in the periodic table with atomic numbers 57 through 71.

Laser: A device for making very intense light of one very specific color that is intensified many times over.

Liquid air: Air that has been cooled to a very low temperature.

Luminescence: The property of giving off light without giving off heat.

Machining: The bending, cutting, and shaping of a metal by mechanical means.

"Magic number": The number of protons and/or neutrons in an atom that tend to make the atom stable (not radioactive).

Magnetic field: The space around an electric current or a magnet in which a magnetic force can be observed.

Malleable: Capable of being hammered into thin sheets.

Metalloid: An element that has characteristics of both metals and non-metals.

Metallurgy: The art and science of working with metals.

Metals: Elements that have a shiny surface, are good conductors of heat and electricity, and can be melted, hammered into thin sheets, and drawn into thin wires.

Micronutrient: A substance needed in very small amounts to maintain good health.

Misch metal: A metal that contains different rare earth elements and has the unusual property of giving off a spark when struck.

Mohs scale: A way of expressing the hardness of a material.

Mordant: A material that helps a dye stick to cloth.

Nanotubes: Long, thin, and extremely tiny tubes.

Native: Not combined with any other element.

Neutron radiography: A technique that uses neutrons to study the internal composition of material.

Nickel allergy: A health condition caused by exposure to nickel metal.

Nitrogen fixation: The process of converting nitrogen as an element to a compound that contains nitrogen.

Noble gases: Elements in Group 18 (VIIIA) of the periodic table.

Non-metals: Elements that do not have the properties of metals.

Nuclear fission: A process in which neutrons collide with the nucleus of a plutonium or uranium atom, causing it to split apart with the release of very large amounts of energy.

Nuclear reactor: A device in which nuclear reactions occur.

Optical fiber: A thin strand of glass through which light passes; the light carries a message through a telephone wire.

Ore: A mineral compound that is mined for one of the elements it contains, usually a metal element.

Organic chemistry: The study of the carbon compounds.

Oxidizing agent: A chemical substance that gives up or takes on electrons from another substance.

Ozone: A form of oxygen that filters out harmful radiation from the sun.

Ozone layer: The layer of ozone that shields Earth from harmful ultraviolet radiation from the sun.

Particle accelerator ("atom smasher"): A device used to cause small particles, such as protons, to move at very high speeds.

Periodic law: A way of organizing the chemical elements to show how they are related to each other.

Periodic table: A chart that shows how chemical elements are related to each other.

Phosphor: A material that gives off light when struck by electrons.

Photosynthesis: The process by which plants convert carbon dioxide and water to carbohydrates (starches and sugars).

Platinum family: A group of elements that occur close to platinum in the periodic table and with platinum in the Earth's surface.

Polymerization: The process by which many thousands of individual tetrafluoroethlylene (TFE) molecules join together to make one very large molecule.

Potash: A potassium compound that forms when wood burns.

Precious metal: A metal that is rare, desirable, and, therefore, expensive.

Proteins: Compounds that are vital to the building and growth of cells.

Pyrophoric: Gives off sparks when scratched.

R

Radiation: Energy transmitted in the form of electromagnetic waves or subatomic particles.

Radioactive isotope: An isotope that breaks apart and gives off some form of radiation.

Radioactive tracer: An isotope whose movement in the body can be followed because of the radiation it gives off.

Radioactivity: The process by which an isotope element breaks down and gives off some form of radiation.

Rare earth elements: Elements in the Lanthanoid family.

Reactive: Combines with other substances relatively easily.

Refractory: A material that can withstand very high temperatures and reflects heat back away from itself.

Rodenticide: A poison used to kill rats and mice.

Rusting: A process by which a metal combines with oxygen.

S

Salt dome: A large mass of salt found underground.

Semiconductor: A material that conducts an electric current, but not nearly as well as metals.

Serendipity: Discovering something of value when not seeking it; for example, making a discovery by chance or accident.

Silver plating: A process by which a very thin layer of silver metal is laid down on top of another metal.

Slag: A mixture of materials that separates from a metal during its purification and floats on top of the molten metal.

Slurry: A soup-like mixture of crushed ore and water.

Solder: An alloy that can be melted and then used to join two metals to each other.

Spectra: The lines produced when chemical elements are heated.

Spectroscope: A device for analyzing the light produced when an element is heated.

Spectroscopy: The process of analyzing light produced when an element is heated.

Spectrum (plural: spectra): The pattern of light given off by a glowing object, such as a star.

Stable: Not likely to react with other materials.

Sublimation: The process by which a solid changes directly to a gas when heated, without first changing to a liquid.

Superalloy: An alloy made of iron, cobalt, or nickel that has special properties, such as the ability to withstand high temperatures and attack by oxygen.

Superconductivity: The tendency of an electric current to flow through a material without resistance.

Superconductor: A material that has no resistance to the flow of electricity; once an electrical current begins flowing in the material, it continues to flow forever.

Superheated water: Water that is hotter than its boiling point, but which has not started to boil.

Surface tension: A property of liquids that makes them act like they are covered with a skin.

Tarnishing: Oxidizing; reacting with oxygen in the air.

Tensile: Capable of being stretched without breaking.

Thermocouple: A device for measuring very high temperatures.

Tin cry: A screeching-like sound made when tin metal is bent.

Tin disease: A change that takes place in materials containing tin when the material is cooled to temperatures below 55°F (13°C) for long periods of time, when solid tin turns to a crumbly powder.

Tincture: A solution made by dissolving a substance in alcohol.

Tinplate: A type of metal consisting of a thin protective coating of tin deposited on the outer surface of some other metal.

Toxic: Poisonous.

Trace element: An element that is needed in very small amounts for the proper growth of a plant or animal.

Tracer: A radioactive isotope whose presence in a system can easily be detected.

Transfermium element: Any element with an atomic number greater than 100.

Transistor: A device used to control the flow of electricity in a circuit.

Transition metal: An element in Groups 3 through 12 of the periodic table.

Transuranium element: An element with an atomic number greater than 92.

Ultraviolet (UV) radiation: Electromagnetic radiation (energy) of a wavelength just shorter than the violet (shortest wavelength) end of the visible light spectrum and thus with higher energy than visible light.

Vulcanizing: The process by which soft rubber is converted to a harder, long-lasting product.

Workability: The ability to work with a metal to get it into a desired shape or thickness.

Actinium

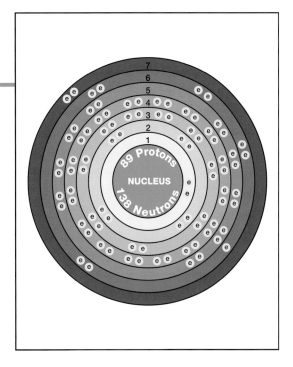

Overview

Actinium is the third element in Row 7 of the periodic table, a chart that shows how the chemical elements are related to each other. Some chemists place it in Group 3 (IIIB), with scandium and yttrium. Other chemists call it the first member of the actinoids. The actinoids are the 15 elements that make up Row 7 of the periodic table after radium. They have atomic numbers from 89 to 103 and are all radioactive. A radioactive atom is unstable and tends to throw off particles and emit energy in order to become stable. Either way of classifying actinium is acceptable to most chemists.

Actinium has chemical properties like those of **lanthanum** (number 57), the element just above it in the periodic table. Actinium is also similar to **radium**, the element just before it (number 88) in Row 7.

Naturally occurring actinium is very rare in Earth's crust. It can be made in the lab by firing neutrons at radium, but it has very few important uses.

Discovery and Naming

Four new elements, all radioactive, were discovered between 1898 and 1900. A radioactive element is one that gives off radiation in the form

Key Facts

Symbol: Ac

Atomic Number: 89

Atomic Mass: [227]

Family: Group 3 (IIIB); transition metal

Pronunciation: ack-TIN-ee-um

of energy or particles and may change into a different element. The first two of these elements—**polonium** and radium—were discovered by French physicists Marie Curie (1867–1934) and Pierre Curie (1859–1906). The third, actinium, was discovered in 1899 by a close friend of the Curies, French chemist André-Louis Debierne (1874–1949). Debierne suggested the name actinium for the new element. The name comes from the Greek words *aktis* or *aktinos,* meaning "beam" or "ray." The fourth element discovered in this series was **radon**, a gas given off during the radioactive decay of some heavier elements. It was found in 1900 by German chemist Friedrich Ernst Dorn (1848–1916).

Actinium was discovered a second time in 1902. German chemist Friedrich O. Giesel (1852–1927) had not heard of Debierne's earlier discovery. Giesel suggested the name emanium, from the word emanation, which means "to give off rays." Debierne's name was adopted, however, because he discovered actinium first.

Physical and Chemical Properties

Only limited information is available about actinium. It is known to be a silver metal with a melting point of 1,920°F (1,050°C) and a boiling point estimated to be about 5,800°F (3,200°C). The element has properties similar to those of lanthanum. Generally speaking, elements in the same column in the periodic table have similar properties. Few compounds of actinium have been produced. Neither the element nor its compounds have any important uses.

Occurrence in Nature

Actinium is found in **uranium** ores. An ore is a mineral mined for the elements it contains. Actinium is produced by the radioactive decay, or breakdown, of uranium and other unstable elements. Actinium can also be artificially produced. When radium is bombarded with neutrons, some of the neutrons become part of the nucleus. This increases the atomic weight and the instability of the radium atom. The unstable radium decays, gives off radiation, and changes to actinium. Actinium metal of 98 percent purity—used for research purposes—can be made by this process.

Isotopes

Thirty-four isotopes of actinium are known, all of which are radioactive. The isotope that occurs in nature is actinium-227. Isotopes are two or more forms of an element. Isotopes differ from each other according to their mass number. The number written to the right of the element's name is the mass number. The mass number represents the number of protons plus neutrons in the nucleus of an atom of the element. The number of protons determines the element, but the number of neutrons in the atom of any one element can vary. Each variation is an isotope. A radioactive isotope is one that breaks apart and gives off some form of radiation.

The half life of actinium-227 is 21.77 years. The half life of a radio-active element is the time it takes for half of a sample of the element to break down. For example, suppose 1.0 gram of actinium-227 is formed by the breakdown of another element. After 21.77 years, only 0.5 gram of actinium-227 would remain. This is known as the half life.

Extraction

Actinium is rarely, if ever, extracted from natural sources.

Uses

There are no practical commercial uses of actinium. Actinium of 98 percent purity is prepared for research studies.

Compounds

The few compounds of actinium that are known are used solely for research purposes.

Health Effects

Like all radioactive materials, actinium is a health hazard. If taken into the body, it tends to be deposited in the bones, where the energy it emits damages or destroys cells. Radiation is known to cause bone cancer and other disorders.

Aluminum

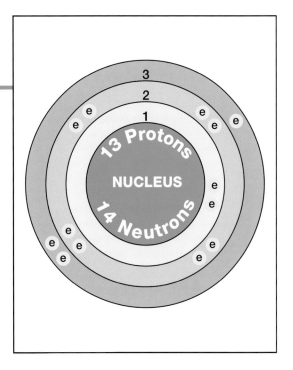

Overview

Aluminum is found in Row 3, Group 13 of the periodic table. The periodic table is a chart that shows how the chemical elements are related to each other. Elements in the same column usually have similar chemical properties. The first element in this group is **boron**. However, boron is very different from all other members of the family. Therefore, group 13 is known as the aluminum family.

Aluminum is the third most abundant element in Earth's crust, exceeded only by **oxygen** and **silicon**. It is second to silicon as the most abundant metallic element. It is somewhat surprising, then, that aluminum was not discovered until relatively late in human history. Aluminum occurs naturally only in compounds, never as a pure metal. Removing aluminum from its compounds is quite difficult. An inexpensive method for producing pure aluminum was not developed until 1886.

Today, aluminum is the most widely used metal in the world after **iron**. It is used in the manufacture of automobiles, packaging materials, electrical equipment, machinery, and building construction. Aluminum is also ideal for beer and soft drink cans and foil because it can be melted and reused, or recycled.

WORDS TO KNOW

Alloy: A mixture of two or more metals with properties different from those of the individual metals.

Ductile: Capable of being drawn into thin wires.

Isotopes: Two or more forms of an element that differ from each other according to their mass number.

Malleable: Capable of being hammered into thin sheets.

Periodic table: A chart that shows how the chemical elements are related to each other.

Radioactive isotope: An isotope that breaks apart and gives off some form of radiation.

Discovery and Naming

Aluminum was named for one its most important compounds, alum. Alum is a compound of **potassium**, aluminum, **sulfur**, and oxygen. Its chemical name is potassium aluminum sulfate, $KAl(SO_4)_2$. In North America, aluminum is spelled with one *i* and is pronounced uh-LOO-min-um. Elsewhere in the world, a second *i* is added—making it aluminium—and the word is pronounced al-yoo-MIN-ee-um.

No one is sure when alum was first used by humans. The ancient Greeks and Romans were familiar with the compound alum. It was mined in early Greece where it was sold to the Turks. The Turks used the compound to make a beautiful dye known as Turkey red. Records indicate that the Romans were using alum as early as the first century BCE.

These early people used alum as an astringent and as a mordant. An astringent is a chemical that causes skin to pull together. Sprinkling alum over a cut causes the skin to close over the cut and start healing. A mordant is used in dyeing cloth. Few natural dyes stick directly to cloth. A mordant bonds to the cloth and the dye bonds to the mordant.

Over time, chemists gradually began to realize that alum might contain a new element. In the mid-1700s, German chemist Andreas Sigismund Marggraf (1709–1782) claimed to have found a new "earth" called alumina in alum. But he was unable to remove a pure metal from alum.

The first person to accomplish that task was Danish chemist and physicist Hans Christian Oersted (1777–1851). Oersted heated a combination of alumina and potassium amalgam. An amalgam is an alloy of a metal and mercury. In this reaction, Oersted produced an aluminum amalgam—aluminum metal in combination with **mercury**. He was unable, however, to separate the aluminum from the mercury.

Pure aluminum metal was finally produced in 1827 by German chemist Friedrich Wöhler (1800–1882). Wöhler used a method perfected by English chemist Sir Humphry Davy (1778–1829), who succeeded in isolating several elements during his lifetime. Wöhler heated a mixture of aluminum chloride and potassium metal. Being more active, the potassium replaces the aluminum, as shown by the following equation:

$$3K + AlCl_3 \rightarrow 3KCl + Al$$

The pure aluminum can then be collected as a gray powder, which must be melted to produce the shiny aluminum that is most familiar to consumers.

After Wöhler's work, it was possible, but very expensive, to produce pure aluminum. It cost so much that there were almost no commercial uses for it.

Before chemists developed inexpensive ways to produce pure aluminum, it was considered a somewhat precious metal. In fact, in 1855, a bar of pure aluminum metal was displayed at the Paris Exposition. It was placed next to the French crown jewels!

A number of chemists realized how important it was to find a less expensive way to prepare aluminum. In 1883, Russian chemist V. A. Tyurin found a less expensive way to produce pure aluminum. He passed an electric current through a molten (melted) mixture of cryolite and **sodium** chloride (ordinary table salt). Cryolite is sodium aluminum fluoride (Na_3AlF_6). Over the next few years, similar methods for isolating aluminum were developed by other chemists in Europe.

The most dramatic breakthrough in aluminum research was made by a college student in the United States. Charles Martin Hall (1863–1914) was a student at Oberlin College in Oberlin, Ohio, when he became interested in the problem of producing aluminum. Using homemade equipment in a woodshed behind his home, he achieved success by passing an electric current through a molten mixture of cryolite and aluminum oxide (Al_2O_3).

Hall's method was far cheaper than any previous method. After his discovery, the price of aluminum fell from about $20/kg ($10/lb) to less than $1/kg (about $.40/lb). Hall's research changed aluminum from a semi-precious metal to one that could be used for many everyday products.

Physical Properties

Aluminum is a silver-like metal with a slightly bluish tint. It has a melting point of 1,220°F (660°C) and a boiling point of 4,221–4,442°F (2,327–2,450°C). The density is 2.708 grams per cubic centimeter. Aluminum is both ductile and malleable. Ductile means capable of being pulled into thin wires. Malleable means capable of being hammered into thin sheets.

Aluminum is an excellent conductor of electricity. **Silver** and **copper** are better conductors than aluminum but are much more expensive. Engineers are looking for ways to use aluminum more often in electrical equipment because of its lower costs.

Chemical Properties

Aluminum has one interesting and very useful property. In moist air, it combines slowly with oxygen to form aluminum oxide:

$$4Al + 3O_2 \rightarrow 2Al_2O_3$$

The aluminum oxide forms a very thin, whitish coating on the aluminum metal. The coating prevents the metal from reacting further with oxygen and protects the metal from further corrosion (rusting). It is easy to see the aluminum oxide on aluminum outdoor furniture and unpainted house siding.

Aluminum is a fairly active metal. It reacts with many hot acids. It also reacts with alkalis. An alkali is a chemical with properties opposite those of an acid. Sodium hydroxide (common lye) and limewater are examples of alkalis. It is unusual for an element to react with *both* acids and alkalis. Such elements are said to be amphoteric.

Aluminum also reacts vigorously with hot water. In powdered form, it catches fire quickly when exposed to a flame.

Occurrence in Nature

The abundance of aluminum in Earth's crust is estimated to be about 8.8 percent. It occurs in many different minerals. Bauxite, a complicated mixture of compounds consisting of aluminum, oxygen, and other elements, is the primary commercial source for aluminum.

As of 2008, large reserves of bauxite were found in Australia, China, Brazil, Guinea, Jamaica, Russia, and Venezuela. Bauxite production statistics for the United States were not reported to protect trade secrets.

According to the U.S. Geological Survey (USGS), in 2008 China led world smelter production of aluminum, followed by Russia, Canada, the United States, Australia, Brazil, and India.

Isotopes

Only one naturally occurring isotope of aluminum exists: aluminum-27. Isotopes are two or more forms of an element. Isotopes differ from each other according to their mass number. The number written to the right of the element's name is the mass number. The mass number represents the number of protons plus neutrons in the nucleus of an atom of the element. The number of protons determines the element, but the number of neutrons in the atom of any one element can vary. Each variation is an isotope.

Aluminum also has 14 radioactive isotopes. A radioactive isotope gives off either energy or subatomic particles in order to reduce the atomic mass and become stable. When the emission produces a change in the number of protons, the atom is no longer the same element. The particles and energy emitted from the nucleus are called radiation. The process of decaying from one element into another is known as radioactive decay.

No radioactive isotope of aluminum has any commercial use.

Extraction

Aluminum production is a two-step process. First, aluminum oxide is separated from bauxite by the Bayer process. In this process, bauxite is mixed with sodium hydroxide (NaOH), which dissolves the aluminum oxide. The other compounds in bauxite are left behind.

The aluminum oxide is then treated with a process similar to the Hall method. There is not enough natural cryolite to make all the aluminum needed, so synthetic (artificial) cryolite is manufactured for this purpose. The chemical reaction is the same with synthetic cryolite as with natural cryolite.

Uses

Aluminum is used as pure metal, in alloys, and in a variety of compounds. An alloy is made by melting and then mixing two or more metals. The mixture has properties different from those of the individual

The aluminum used for beer and soft drink cans can be recycled. IMAGE COPYRIGHT 2009, KATHY BURNS-MILLYARD. USED UNDER LICENSE FROM SHUTTERSTOCK.COM.

metals. Aluminum alloys are classified in numbered series according to the other elements they contain.

The 1000 classification is reserved for alloys of nearly pure aluminum metal. They tend to be less strong than other alloys of aluminum, however. These metals are used in the structural parts of buildings, as decorative trim, in chemical equipment, and as heat reflectors.

The 2000 series are alloys of copper and aluminum. They are very strong, are corrosion (rust) resistant, and can be machined, or worked with, very easily. Some applications of 2000 series aluminum alloys are in truck paneling and structural parts of aircraft.

The 3000 series is made up of alloys of aluminum and **manganese**. These alloys are not as strong as the 2000 series, but they also have good machinability. Alloys in this series are used for cooking utensils, storage tanks, aluminum furniture, highway signs, and roofing.

Alloys in the 4000 series contain silicon. They have low melting points and are used to make solders and to add gray coloring to metal. Solders are low-melting alloys used to join two metals to each other. The 5000, 6000, and 7000 series include alloys consisting of **magnesium**, both magnesium and silicon, and **zinc**, respectively. These alloys are used in ship and boat production, parts for cranes and gun mounts, bridges, structural parts in buildings, automobile parts, and aircraft components.

The largest single use of aluminum in the United States is in the transportation industry (37 percent). Car and truck manufacturers like aluminum and aluminum alloys because they are very strong, yet lightweight. Companies are using more aluminum products in electric cars. These cars must be lightweight in order to conserve battery power. Aluminum producers also plan to make a wider variety of wheels for both cars and trucks.

Another major use of aluminum is in packaging (23 percent). Aluminum foil, beer and soft drink cans, paint tubes, and containers for home products such as aerosol sprays are all made from aluminum.

Aluminum is also used for building and construction (13 percent). Windows and door frames, screens, roofing, and siding, as well as the construction of mobile homes and structural parts of buildings rely on aluminum.

Aluminum is also used in a staggering range of products, including electrical wires and appliances, automobile engines, heating and cooling systems, bridges, vacuum cleaners, kitchen utensils, garden furniture, heavy machinery, and specialized chemical equipment.

Compounds

A relatively small amount of aluminum is used to make a large variety of aluminum compounds. These include:

- aluminum ammonium sulfate (Al(NH$_4$)(SO$_4$)$_2$): mordant, water purification and sewage treatment, paper production, food additive, leather tanning
- aluminum borate (Al$_2$O$_3$ • B$_2$O$_3$): production of glass and ceramics
- aluminum borohydride (Al(BH$_4$)$_3$): additive in jet fuels
- aluminum chloride (AlCl$_3$): paint manufacture, antiperspirant, petroleum refining, production of synthetic rubber
- aluminum fluorosilicate (Al$_2$(SiF$_6$)$_3$): production of synthetic gemstones, glass, and ceramics
- aluminum hydroxide (Al(OH)$_3$): antacid, mordant, water purification, manufacture of glass and ceramics, waterproofing of fabrics
- aluminum phosphate (AlPO$_4$): manufacture of glass, ceramics, pulp and paper products, cosmetics, paints and varnishes, and in making dental cement
- aluminum sulfate, or alum (Al$_2$(SO$_4$)$_3$): manufacture of paper, mordant, fire extinguisher system, water purification and sewage treatment, food additive, fireproofing and fire retardant, and leather tanning

Health Effects

Aluminum has no known function in the human body. There is some debate, however, as to its possible health effects. Some health scientists suspect that aluminum may be associated with Alzheimer's disease. This is a condition that most commonly affects older people, leading to forgetfulness and loss of mental skills. It is still not clear whether aluminum plays any part in Alzheimer's disease.

Breathing aluminum dust may also cause health problems. It may cause a pneumonia-like condition called aluminosis.

Americium

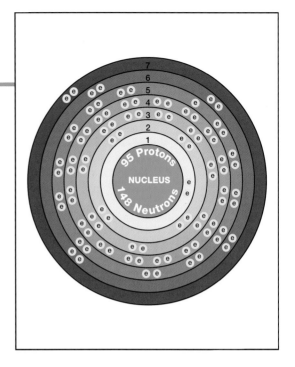

Overview

Americium is called an actinoid or transuranium element. It occurs in Row 7 of the periodic table, a chart that shows how the chemical elements are related to each other. The actinoids are named after element 89, actinium. The term transuranium means "beyond uranium" in the periodic table. **Uranium** has an atomic number of 92. Any element with an atomic number larger than 92, therefore, is called a transuranium element.

Discovery and Naming

Americium was discovered as a by-product of military research during World War II (1939–45). The U.S. government maintained a major research site at the University of Chicago during the war. Work there led to the development of the first atomic bomb. During that research, a team from the University of California, consisting of Glenn Seaborg (1912–1999), Albert Ghiorso (1915–), Ralph A. James, and Leon O. Morgan, discovered a new element, which would eventually be named americium.

Americium was first produced in a nuclear reactor. A nuclear reactor is a device in which elements are bombarded by neutrons, sometimes

Key Facts

Symbol: Am

Atomic Number: 95

Atomic Mass: [243]

Family: Actinoid; transuranium element

Pronunciation: am-uh-REE-see-um

WORDS TO KNOW

Actinoid family: Elements with atomic numbers 89 through 103.

Alpha radiation: A form of radiation that consists of very fast moving alpha particles and helium atoms without their electrons.

Half life: The time it takes for half of a sample of a radioactive element to break down.

Isotopes: Two or more forms of an element that differ from each other according to their mass number.

Nuclear reactor: A device in which nuclear reactions occur.

Periodic table: A chart that shows how the chemical elements are related to each other.

Radioactive: Having a tendency to give off radiation.

Transuranium element: An element with an atomic number greater than 92.

forming new elements. The name, americium, in honor of the American continent, was chosen because the new element occurs just below **europium** (named for Europe) in the periodic table.

Physical Properties

Enough americium has been produced to determine a few of its properties. It is a silvery-white metal with a melting point of about 2,150°F (1,175°C) and a density of about 13.6 grams per cubic centimeter. A number of its compounds have been produced and studied, but only one isotope has considerable practical use outside the laboratory.

Occurrence in Nature

All of the transuranium elements, including americium, are synthetically produced. None exist in nature.

Isotopes

Americium has 25 isotopes, all of which are radioactive. The most stable is americium-243. Isotopes are two or more forms of an element. Isotopes differ from each other according to their mass number. The number written to the right of the element's name is the mass number. The mass number represents the number of protons plus neutrons in the nucleus of an atom of the element. The number of protons determines the element, but the number of neutrons in the atom of any one element can vary.

Each variation is an isotope. A radioactive isotope is one that breaks apart and gives off some form of radiation.

The half life of a radioactive element is the time it takes for half of a sample of the element to break down. The half life of americium-243 is 7,370 years. For example, suppose a laboratory made 10 grams of americium-243. At the end of 7,370 years (one half life), only half would remain. The other half would have changed into a new element.

Extraction

Americium does not occur naturally.

Uses

Americium-241 is the only isotope of americium of any practical interest. When it decays, it gives off both alpha rays and gamma rays. Alpha rays do not travel very far in air, but gamma rays are very penetrating, much like X rays. The gamma rays from americium-241 are used in portable X-ray machines that can, for example, be taken into oil fields to help determine where new wells should be dug.

Americium-241 is also used to measure the thickness of materials. For instance, a small piece of americium-241 can be placed above a conveyor belt carrying newly made glass. A Geiger counter, a device for counting alpha radiation, is placed below the conveyor belt. If the glass is always the same thickness, the same amount of alpha radiation gets through to the detector. If the glass is thicker than normal, less alpha radiation gets through. If the glass is thinner than normal, more radiation gets through. The detector will register if the glass being produced is too thick or too thin.

One of the simplest and cheapest safety devices found in homes and other buildings is a battery-operated smoke detector. And americium is an important part of it. A small piece of americium oxide made with the americium-241 isotope is sealed inside the smoke detector. The americium-241 gives off alpha particles. The alpha particles strike air molecules, causing them to break apart. The pieces formed in this process—ions—are electrically charged.

The electrically charged ions help carry a current from one side of the detector to the other. The current continues to flow as long as nothing other than air is inside the detector. If smoke enters the detector, the smoke particles absorb some of the alpha particles so that the current is interrupted. When this happens, a buzzer or other sound is set off.

Common smoke detectors include an americium oxide made with the americium-241 isotope. IMAGE COPYRIGHT 2009, DANNY E. HOOKS. USED UNDER LICENSE FROM SHUTTERSTOCK.COM.

Compounds

There are no commercial uses of americium compounds.

Health Effects

Americium is an extremely toxic element. If swallowed, it is deposited in the bones. There, the radiation it gives off kills or damages cells, causing cancer. People are normally in no danger from smoke detectors containing americium-241. (Indeed, countless lives are saved each year because of smoke detectors.) The amount of this isotope in a smoke detector is very small. One gram of americium oxide made with americium-241 will make 5,000 smoke detectors.

Antimony

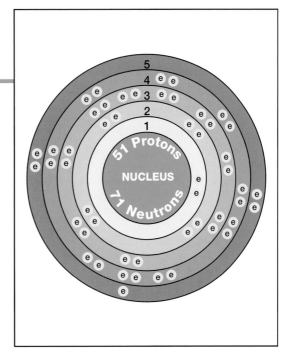

Key Facts

Symbol: Sb

Atomic Number: 51

Atomic Mass: 121.760

Family: Group 15 (VA); nitrogen

Pronunciation: AN-ti-moh-nee

Overview

Antimony compounds have been used by humans for centuries. Women of ancient Egypt used stibic stone, antimony sulfide (Sb_2S_3), to darken the skin around their eyes. Antimony was also used in making colored glazes for beads and glassware. The chemical symbol for antimony was taken from the ancient name for the element, stibium. Not recognized as a chemical element until the Middle Ages, antimony became a common material used by alchemists.

Alchemy was a kind of pre-science that existed from about 500 BCE to about the end of the 16th century. Alchemists wanted to find a way of changing **lead**, **iron**, and other metals into **gold**. They also wanted to find a way of having eternal life. Alchemy contained too much magic and mysticism to be a real science, but alchemists developed a number of techniques and produced many new materials that were later found to be useful in modern chemistry. Antimony was one of these materials.

Antimony is a metalloid. A metalloid is an element that has characteristics of both metals and non-metals. The metalloids can be found on either side of the staircase line on the right side of the periodic table (with the exception of **aluminum**, which is not considered a metalloid).

19

WORDS TO KNOW

Alchemy: A kind of pre-science that existed from about 500 BCE to about the end of the 16th century.

Alloy: A mixture of two or more metals with properties different from those of the individual metals.

Aqua regia: A mixture of hydrochloric and nitric acids that often reacts with materials that do not react with either acid separately.

Catalyst: A substance used to speed up a chemical reaction without undergoing any change itself.

Isotopes: Two or more forms of an element that differ from each other according to their mass number.

Metalloid: An element that has characteristics of both metals and non-metals.

Periodic table: A chart that shows how the chemical elements are related to each other.

Radioactive isotope: An isotope that breaks apart and gives off some form of radiation.

Solder: An alloy that can be melted and then used to join two metals to each other.

Toxic: Poisonous.

Tracer: A radioactive isotope whose presence in a system can be easily detected.

Antimony is primarily used in alloys, ceramics and glass, plastics, and flame retardant materials. Flame retardant materials do not burn with an open flame. Instead, they smolder or do not burn at all.

Discovery and Naming

Compounds of antimony were known to ancient cultures. They have been found, for example, in the colored glazes used on beads, vases, and other glassware. But these compounds were not widely used until the Middle Ages when they became popular among alchemists. They thought that antimony could be used to convert lead into gold. It was during this period that records about the properties of antimony begin to appear.

The element was probably first named by Roman scholar Pliny (23–79 CE), who called it stibium. Muslim alchemist Abu Musa Jabir Ibn Hayyan (c. 721–c. 815) probably first called it antimony—*anti* ("not") and *monos* ("alone"). The name comes from the fact that antimony does not occur alone in nature.

Alchemists used secret codes to write about much of their work, so modern scholars do not know a great deal about how antimony was used. The first detailed reports about antimony were published in 1707 when French chemist Nicolas Lemery (1645–1715) published his famous book, *Treatise on Antimony.*

Antimony samples. ©RUSS LAPPA/SCIENCE SOURCE, NATIONAL AUDUBON SOCIETY COLLECTION/PHOTO RESEARCHERS, INC.

Physical Properties

Antimony is a silvery-white, shiny element that looks like a metal. It has a scaly surface and is hard and brittle like a non-metal. It can also be prepared as a black powder with a shiny brilliance to it.

The melting point of antimony is 1,170°F (630°C) and its boiling point is 2,980°F (1,635°C). It is a relatively soft material that can be scratched by glass. Its density is 6.68 grams per cubic centimeter.

Chemical Properties

Antimony is a moderately active element. It does not combine with **oxygen** in the air at room temperature. It also does not react with cold water or with most cold acids. It does dissolve in some hot acids, however, and in aqua regia (a mixture of hydrochloric and nitric acids). It often reacts with materials that do not react with either acid separately.

Occurrence in Nature

Antimony is rarely found in its native (as an element) state. Instead, it usually occurs as a compound. The most common minerals of antimony are stibnite, tetrahedrite, bournonite, boulangerite, and jamesonite. In most of these minerals, antimony is combined with **sulfur** to produce some form of antimony sulfide (Sb_2S_3).

In 2008, the largest producers of antimony were China, Bolivia, South Africa, Russia, and Tajikistan. The United States produced some antimony concentrate at a mine in Nevada and some antimony metal and oxide in Montana. According to the U.S. Geological Survey (USGS), other states with antimony resources include Alaska and Idaho.

The abundance of antimony is estimated to be about 0.2 parts per million, placing it in the bottom fifth among the chemical elements found in Earth's crust. It is more abundant than **silver** or **mercury**, but less abundant than **iodine**.

Isotopes

There are two naturally occurring isotopes of antimony: antimony-121 and antimony-123. Isotopes are two or more forms of an element. Isotopes differ from each other according to their mass number. The number written to the right of the element's name is the mass number. The mass number represents the number of protons plus neutrons in the nucleus of an atom of the element. The number of protons determines the element, but the number of neutrons in the atom of any one element can vary. Each variation is an isotope.

Forty-three radioactive isotopes of antimony are also known. A radioactive isotope is one that breaks apart and gives off some form of radiation. Radioactive isotopes are produced when very small particles are fired at atoms. These particles stick in the atoms and make them radioactive.

Two of antimony's radioactive isotopes are used commercially as tracers. These isotopes are antimony-124 and antimony-125. A tracer is an isotope injected into a living or non-living system. The movement of the isotope can then be followed as it moves through the system. For example, a small amount of antimony-124 could be injected into an oil pipeline. The presence of the isotope can be detected by means of an instrument held above the pipeline. The radiation given off by the isotope causes a light to flash or a sound to occur in the instrument.

The movement of the isotope through the pipeline can be followed in this way. If the pipeline has a leak, the tracer will escape from it. Its movement through the soil can be detected.

Extraction

Antimony can be recovered from stibnite with hot iron:

$$2Fe + Sb_2S_3 \rightarrow Fe_2S_3 + 2Sb$$

Some of the antimony produced in the United States is recycled from old lead storage batteries used in cars and trucks.

Uses

Antimony is used to make alloys with a number of different metals. An alloy is made by melting and mixing two or more metals. The properties of the mixture are different than those of the individual metals. One of the most common of these alloys is one made with lead. Lead-antimony alloys are used for solder, ammunition, fishing tackle, covering for electrical cables, alloys that melt at low temperatures, and batteries.

The manufacture of lead storage batteries, like the ones used in cars and trucks, account for about one-fifth of all the antimony used each year. A small amount of antimony is also used in making transistors, which are found in such consumer electrical devices as computer games, pocket calculators, and portable stereos. A transistor is a solid-state (using special properties of solids, rather than electron tubes) electronic device used to control the flow of an electric current.

Other minor uses of antimony include the manufacture of glass and ceramics and the production of plastics. In glass and ceramics, a small amount of antimony insures that the final product will be clear and colorless. In the production of plastics, antimony is used as a catalyst. A catalyst is a substance that increases the rate of a chemical reaction. The catalyst does not undergo any change itself during the reaction.

Compounds

Commercially, antimony trioxide (Sb_2O_3) is the most important compound of antimony. It is used primarily in the manufacture of flame-retardant materials. It is usually sprayed on or added to a fabric to make it flame retardant.

Health Effects

Antimony and its compounds are dangerous to human health. In low levels, these materials can irritate the eyes and lungs. They may also cause stomach pain, diarrhea, vomiting, and stomach ulcers. At higher doses, antimony and its compounds can cause lung, heart, liver, and kidney damage. At very high doses, they can cause death.

Argon

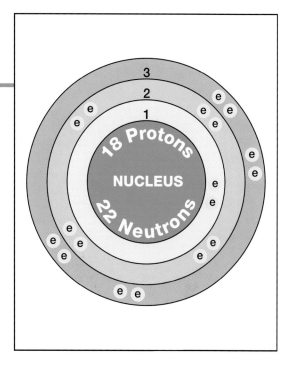

Overview

Argon is a noble gas. The noble gases are the six elements in Group 18 (VIIIA) of the periodic table. The periodic table is a chart that shows how the chemical elements are related to each other. The noble gases are sometimes called inert gases because Group 18 (VIIIA) elements react with very few other elements. In fact, no compound of argon had been made until 2000. In that year, researchers at the University of Helsinki, in Finland, made the first compound of argon by reacting the element with hydrogen fluoride. The compound formed was argon hydrofluoride (HArF).

Argon was discovered in 1894 by English chemist John William Strutt, most commonly known as Lord Rayleigh (1842–1919), and Scottish chemist William Ramsay (1852–1916). It was the first of the noble gases to be isolated.

Rayleigh and Ramsay discovered argon by the fractional distillation of liquid air. Fractional distillation is the process of letting liquid air slowly warm up. As the air warms, different elements change from a liquid back to a gas. The portion of air that changes back to a gas at −302.55°F (−185.86°C) is argon.

Key Facts

Symbol: Ar

Atomic Number: 18

Atomic Mass: 39.948

Family: Group 18 (VIIIA); noble gas

Pronunciation: AR-gon

WORDS TO KNOW

Inert: Not very active.

Isotopes: Two or more forms of an element that differ from each other according to their mass number.

Laser: A device for making very intense light of one very specific color that is intensified many times over.

Noble gas: An element in Group 18 (VIIIA) of the periodic table.

Periodic table: A chart that shows how the chemical elements are related to each other.

Radioactive isotope: An isotope that breaks apart and gives off some form of radiation.

Argon is used to provide an inert blanket for certain industrial operations. An inert blanket of gas prevents any chemicals in the operation from reacting with **oxygen** and other substances present in air. Argon is also used in making "neon" lamps and in lasers.

Discovery and Naming

Argon was discovered in 1894. However, English scientist Henry Cavendish (1731–1810) had predicted the existence of argon many years earlier. When Cavendish removed oxygen and **nitrogen** from air, he found that a very small amount of gas remained. He guessed that another element was in the air, but he was unable to identify what it was.

When Ramsay repeated Cavendish's experiments in the 1890s, he, too, found a tiny amount of unidentified gas in the air. But Ramsay had an advantage over Cavendish: he could use spectroscopy, which did not exist in Cavendish's time. Spectroscopy is the process of analyzing light produced when an element is heated. The spectrum (plural: spectra) of an element consists of a series of colored lines and is different for every element.

Ramsay studied the spectrum of the unidentified gas. He found a series of lines that did not belong to any other element. He was convinced that he had found a new element. Meanwhile, Rayleigh was doing similar work and made his discovery at about the same time Ramsay did. The two scientists decided to make their announcement together. The name argon comes from the Greek word *argos,* "the lazy one." The name is based on the fact that argon does not react with other elements or compounds.

The discovery of argon created a problem for chemists. It was the first noble gas to be discovered. Where should it go in the periodic table?

At the time, the table ended with Group 17 (VIIA) at the right. Ramsay suggested that the periodic table might have to be extended. He proposed adding a whole new group to the table. That group would be placed to the right of Group 17 (VIIA).

Ramsay's suggestion was accepted, but it created an interesting new problem for chemists. If there was a new group in the periodic table, where were the other elements that belonged in the group?

Fortunately, chemists had a good idea what these missing elements might look like. All of the elements in a single group are very much like each other. Chemists looked for more inactive gases. Within the next five years, they had found the remaining members of the group: **helium**, **krypton**, **neon**, **radon**, and **xenon**.

The symbol *A* was used for argon until the 1950s when chemists agreed to use the two letter symbol *Ar* for the element.

Physical Properties

Argon is a colorless, odorless, tasteless gas. Its density is 1.784 grams per liter. The density of air, for comparison, is about 1.29 grams per liter. Argon changes from a gas to a liquid at −302.55°F (−185.86°C). Then it changes from a liquid to a solid at −308.7°F (−189.3°C).

Chemical Properties

Argon is chemically inactive. On rare occasions, and under extreme conditions, it forms weak, compound-like structures.

Occurrence in Nature

Argon is the third most abundant gas in Earth's atmosphere, following nitrogen and oxygen. Its abundance is about 0.93 percent. It is also found in Earth's crust to the extent of about 4 parts per million.

Extraction

Argon can be produced from liquid air by fractional distillation. It can also be produced by heating nitrogen gas from the atmosphere with hot **magnesium** or **calcium**. The magnesium or calcium combines with nitrogen to form a nitride:

$$3Mg + N_2 \rightarrow Mg_3N_2$$

A little argon always occurs as an impurity with nitrogen. It remains behind because it does not react with magnesium or calcium.

Argon also occurs in wells with natural gas. When the natural gas is purified, some argon can be recovered as a by-product.

Isotopes

Three isotopes of argon exist naturally. They are argon-36, argon-38, and argon-40. Isotopes are two or more forms of an element. Isotopes differ from each other according to their mass number. The number written to the right of the element's name is the mass number. The mass number represents the number of protons plus neutrons in the nucleus of an atom of the element. The number of protons determines the element, but the number of neutrons in the atom of any one element can vary. Each variation is an isotope.

Fifteen radioactive isotopes of argon are known also. A radioactive isotope is one that breaks apart and gives off some form of radiation. Radioactive isotopes are produced when very small particles are fired at atoms. These particles stick in the atoms and make them radioactive.

No radioactive isotopes of argon have any practical application. One non-radioactive isotope is used, however, to find the age of very old rocks. This method of dating rocks is described in the **potassium** entry.

Uses

Argon is used in situations where materials need to be protected from oxygen or other gases. One example is an incandescent lightbulb, which consists of a metal wire inside a clear glass bulb. An electric current passes through the wire, causing it to get very hot and give off light.

Oxygen will combine with the hot metal very easily, forming a compound of the metal and oxygen. This compound will not conduct an electric current very well, thereby causing the lightbulb to stop giving off light.

Argon, however, is used to prevent this from happening. Because argon is inert, it will not react with the hot wire, leaving the metal hot for very long periods of time. The lightbulb will stop giving off light only when the metal breaks. Then it can no longer carry an electric current.

Argon is also used in welding. Welding is the process by which two metals are joined to each other. In most cases, the two metals are heated to very high temperatures. As they get hot, they melt together.

However, as the metals get hot, they begin to react with oxygen. In this reaction, a compound of metal and oxygen is formed. It becomes very difficult to join the two metals if they have formed compounds, but introducing argon into the welding environment prevents the metals from reacting with oxygen.

Argon is also used in compact fluorescent lightbulbs (CFLs), argon lasers, and argon-dye lasers. A laser is a device that produces very bright light of a single color (frequency). An argon laser is used to treat skin conditions. The laser shines a blue-green light on the affected area of the skin. The energy from the laser is absorbed by hemoglobin and converted to heat. (Hemoglobin is the protein pigment in red blood cells. It transports oxygen to the tissues and carbon dioxide from them.) The blood vessels are damaged, but then sealed, prompting them to decompose and be reabsorbed into the body. Unwanted growths are flattened and dark spots are lightened, with only a small risk of scarring.

An argon-dye laser is used in eye surgery. The color of light produced by the laser can be adjusted with high precision. It can be made to produce light ranging across the green-to-blue color range. Each shade of green or blue has a slightly different frequency. It can penetrate more or less deeply in the eye. The laser can be adjusted to treat a very specific

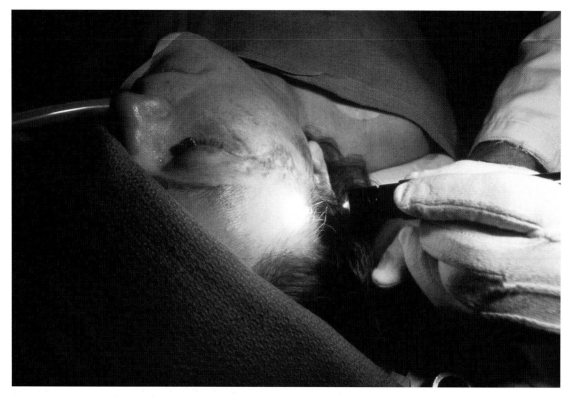

Argon is used in lasers, such as the one shown here removing a birthmark. © ALEXANDER TSIARAS/PHOTO RESEARCHERS, INC.

part of the eye. The argon dye laser is used to treat tumors, damaged blood vessels, conditions involving the retina, and other kinds of eye problems.

Compounds

Only one compound of argon has been produced, argon hydrofluoride (HArF).

Health Effects

Argon is not known to have any positive or negative effects on the health of plants or animals.

Arsenic

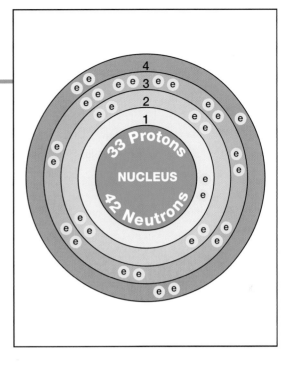

Overview

Arsenic compounds have been known since at least the days of Ancient Greece and Rome (about 2,000 years ago). The compounds were used by physicians and poisoners. The compound most often used for both purposes was arsenic sulfide (As_2S_3).

Arsenic was first recognized as an element by alchemists. Alchemy was a kind of pre-science that existed from about 500 BCE to about the end of the 16th century. People who studied alchemy—alchemists—wanted to find a way of changing **lead**, **iron**, and other metals into **gold**. They were also looking for a way to have eternal life. Alchemy contained too much magic and mysticism to be a real science, but alchemists developed a number of techniques and produced many new materials that were later found to be useful in modern chemistry.

A small amount of arsenic is used in alloys. An alloy is made by melting and then mixing two or more metals. The mixture has properties different from those of individual metals. The most important use of arsenic in the United States is in wood preservatives.

Key Facts

Symbol: As

Atomic Number: 33

Atomic Mass: 74.92160

Family: Group 15 (VA); nitrogen

Pronunciation: AR-se-nick

WORDS TO KNOW

Alchemy: A kind of pre-science that existed from about 500 BCE to about the end of the 16th century.

Allotropes: Forms of an element with different physical and chemical properties.

Alloy: A mixture of two or more metals with properties different from those of the individual metals.

Isotopes: Two or more forms of an element that differ from each other according to their mass number.

Metalloid: An element that has properties of both metals and non-metals.

Periodic table: A chart that shows how the chemical elements are related to each other.

Radioactive isotope: An isotope that breaks apart and gives off some form of radiation.

Sublimation: The process by which a solid changes directly to a gas when heated, without first changing to a liquid.

Toxic: Poisonous.

Discovery and Naming

Arsenic can be produced from its ores very easily, so many early crafts-people may have seen the element without realizing what it was. Since arsenic is somewhat similar to **mercury**, early scholars probably confused the two elements with each other.

Credit for the actual discovery of arsenic often goes to alchemist Albert the Great (Albertus Magnus, 1193–1280). He heated a common compound of arsenic, orpiment (As_2S_3), with soap. Nearly pure arsenic was formed in the process.

By the mid-17th century, arsenic was well known as an element. Textbooks from the period often listed methods by which the element could be made from its compounds.

Physical Properties

Arsenic occurs in two allotropic forms. Allotropes are forms of an element with different physical and chemical properties. The more common form of arsenic is a shiny, gray, brittle, metallic-looking solid. The less common form is a yellow crystalline solid. It is produced when vapors of arsenic are cooled suddenly.

When heated, arsenic does not melt, as most solids do. Instead, it changes directly into a vapor (gas). This process is known as sublimation. However, under high pressure, arsenic can be forced to melt at about

1,500°F (814°C). Arsenic has a density of 5.72 grams per cubic centimeter.

Chemical Properties

Arsenic is a metalloid. A metalloid is an element that has properties of both metals and non-metals. Metalloids occur in the periodic table on either side of the staircase line that starts between **boron** and **aluminum**.

When heated in air, arsenic combines with **oxygen** to form arsenic oxide (As_2O_3). A blue flame is produced, and arsenic oxide can be identified by its distinctive garlic-like odor.

Arsenic combines with oxygen more slowly at room temperatures. The thin coating of arsenic oxide that forms on the element prevents it from reacting further. Arsenic does not dissolve in water or most cold acids. It does react with some hot acids to form arsenous acid (H_3AsO_3) or arsenic acid (H_3AsO_4).

Occurrence in Nature

Arsenic rarely occurs as a pure element. It is usually found as a compound. The most common ores of arsenic are arsenopyrite ($FeAsS$), orpiment (As_2S_3), and realgar (As_4S_4). These compounds are obtained as a by-product of the mining and purification of **silver** metal.

The abundance of arsenic in Earth's crust is thought to be about 5 parts per million. That places it among the bottom third of the elements in abundance in Earth's crust.

In 2008, the world's largest producers of arsenic trioxide were China, Chile, Morroco, and Peru. According to the U.S. Geological Survey (USGS), the United States has not produced arsenic compounds or metal since 1985. About $7 million of arsenic was used in the United States in 2008.

Isotopes

One naturally occurring isotope of arsenic exists: arsenic-75. Isotopes are two or more forms of an element. Isotopes differ from each other according to their mass number. The number written to the right of the element's name is the mass number. The mass number represents the number of protons plus neutrons in the nucleus of an atom of the element. The number of protons determines the element, but the number of neutrons in the atom of any one element can vary. Each variation is an isotope.

Twenty-four radioactive isotopes of arsenic are known also. A radioactive isotope is one that breaks apart and gives off some form of radiation. Radioactive isotopes are produced when very small particles are fired at atoms. These particles stick in the atoms and make them radioactive.

None of the isotopes of arsenic have any important commercial use.

Extraction

The process of recovering arsenic from its ores is a common one used with metals. The ore is first roasted (heated in air) to convert arsenic sulfide to arsenic oxide. The arsenic oxide is then heated with charcoal (pure carbon). The carbon reacts with the oxygen in arsenic oxide, leaving behind pure arsenic:

$$2As_2O_3 + 3C \rightarrow 3CO_2 + 4As$$

Uses

Arsenic is used most commonly in the form of its compounds. A much smaller amount of the element itself is used in alloys. For example, certain parts of lead storage batteries used in cars and trucks contain alloys of lead and arsenic. Arsenic has also been used to make lead shot in the past. The amount of arsenic used in these applications is likely to continue to decrease. It is too easy for arsenic to get into the environment from such applications.

Some arsenic is used in the electronics industry. It is added to **germanium** and **silicon** to make transistors. A compound of arsenic, **gallium** arsenide (GaAs), is also used to make light-emitting diodes (LEDs). LEDs produce the lighted numbers in hand-held calculators, clocks, watches, and a number of other electronic devices.

Compounds

Arsenic has a fascinating history as a healer and killer. Early physicians, such as Hippocrates (c. 460–370 BCE) and Paracelsus (1493–1541), recommended arsenic for the treatment of some diseases. In more recent times, compounds of arsenic have been used to treat a variety of diseases, including syphilis and various tropical diseases.

Arsenic has a special place in the history of modern medicine. In 1910, German biologist Paul Ehrlich (1854–1915) invented the first drug that would cure syphilis, a sexually transmitted disease. This drug, called salvarsan, is a compound of arsenic. Its chemical name is arsphenamine.

Compounds of arsenic have long been used for less happy purposes. Especially during the Middle Ages, they were a popular form of committing murder. At the time, it was difficult to detect the presence of arsenic in the body. A person murdered with arsenic was often thought to have died of pneumonia.

The toxic properties of arsenic compounds made them useful as rat poison. However, they are seldom used for this purpose today. Safer compounds are used that do not present a threat to humans, pets, and the environment.

For many years, the most important application of arsenic was the preservation of wood. When arsenic compounds are added to wood, they kill the insects that attack and eat the wood. In the first years of the 21st century, more than 90 percent of all the arsenic produced in the

For years, historians wondered whether Zachary Taylor, the 12th president of the United States, was poisoned.

United States was used for this purpose. The most common compound used for this purpose was chromated copper arsenate (CCA). Wood preserved with CCA is referred to as pressure-treated wood. Wood treated with CCA is now recognized as a health hazard. Many authorities believe that humans and other animals exposed to pressure-treated wood may develop health problems because of arsenic present in the wood. For this reason, the U.S. Environmental Protection Agency (EPA) issued a ban on the use of CCA for treating wood, effective December 31, 2003. Pressure-treated wood may no longer be used for residential construction, although its use for industrial production is still permitted.

Health Effects

Arsenic and its compounds are toxic to animals. In low doses, arsenic produces nausea, vomiting, and diarrhea. In larger doses, it causes abnormal heart beat, damage to blood vessels, and a feeling of "pins and needles" in hands and feet. Small corns or warts may begin to develop on the palms of the hands and the soles of the feet. Direct contact with the skin can cause redness and swelling.

Long term exposure to arsenic and its compounds can cause cancer. Inhalation can result in lung cancer. If arsenic is swallowed, cancer is likely to develop in the bladder, kidneys, liver, and lungs. In large doses, arsenic and its compounds can cause death. In fact, for years, historians wondered whether the 12th president of the United States, Zachary Taylor (1784–1850), had been murdered—poisoned by arsenic.

On July 9, 1850, Taylor died in office. He had served as president for a little more than 16 months. The cause of death was widely reported as gastroenteritis (an inflammation in the stomach and intestines). He had become sick after eating a mixture of cherries and buttermilk. But many wondered whether Taylor's enemies had actually poisoned the former war hero.

On June 17, 1991, Taylor's remains were exhumed (removed from his grave) from a cemetery in Louisville, Kentucky. The late president's descendents agreed with historians that the possibility of poisoning existed. Samples of Taylor's hair and fingernails were taken to Oak Ridge National Laboratory, in Oak Ridge, Tennessee, for analysis.

Scientists used a process that measured the amount of arsenic in the tissue samples. Most human bodies *do* contain traces of arsenic. So the key issue was whether there would be more arsenic in the tissue samples than would be normal for someone who had been dead for 141 years. If there were, that would mean Taylor was probably poisoned; if not, death by natural causes was more likely.

The Kentucky medical examiner came to a conclusion. He said the amount of arsenic found in Taylor's samples was several hundred times less than what could be expected had the president been poisoned by arsenic. So although some people still wonder whether Taylor was poisoned, arsenic was not the chemical element used.

Astatine

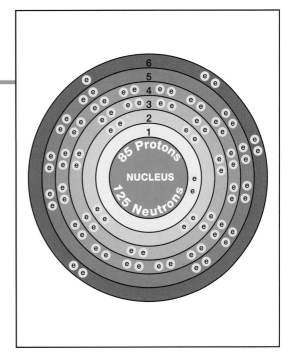

Overview

Astatine is a member of the halogen family, elements in Group 17 (VIIA) of the periodic table. It is one of the rarest elements in the universe. Scientists believe that no more than 25 grams exist on Earth's surface. All isotopes of astatine are radioactive and decay into other elements. For this reason, the element's properties are difficult to study. What *is* known is that it has properties similar to those of the other halogens—**fluorine**, **chlorine**, **bromine**, and **iodine**. Because it is so rare, it has essentially no uses.

Discovery and Naming

The periodic table is a chart that shows how the chemical elements are related to each other. The periodic table was first constructed by Russian chemist Dmitri Mendeleev (1834–1907) in the early 1870s.

Mendeleev's periodic table contained some empty boxes. At first, no one was sure what these empty boxes meant. By the early 1900s, however, chemists had decided that the empty boxes must be spaces for elements that had not yet been discovered. A search began for elements to fill the half dozen or so boxes that still remained in the periodic table.

Key Facts

Symbol: At

Atomic Number: 85

Atomic Mass: [210]

Family: Group 17 (VIIA); halogen

Pronunciation: AS-tuh-teen

WORDS TO KNOW

Cyclotron: A particle accelerator, or "atom smasher," in which small particles, such as protons, are made to travel very fast and then collide with atoms, causing the atoms to break apart.

Half life: The time it takes for half of a sample of a radioactive element to break down.

Halogen: One of the elements in Group 17 (VIIA) of the periodic table.

Isotopes: Two or more forms of an element that differ from each other according to their mass number.

Periodic table: A chart that shows how the chemical elements are related to each other.

Radioactivity: The tendency for an isotope to break apart and give off some form of radiation.

Two of the most troubling empty boxes were elements 85 and 87. During the first third of the 20th century, chemists worked very hard to find these two missing elements. Along the way, a number of incorrect answers were proposed. For example, American chemist Fred Allison (1882–1974) announced in 1931 that he had discovered elements 85 and 87. He proposed the names virginium and alabamine for these two elements. (Allison was born in Virginia and worked at the Alabama Polytechnic Institute.) Unfortunately for Allison, other chemists could not repeat his experiments successfully. They decided his results must have been incorrect.

In 1940, three chemists working at the University of California at Berkeley found evidence of element 85. Dale R. Corson, Kenneth R. Mackenzie, and Emilio Segrè (1905–1989) found evidence of element 85 at the end of an experiment they were conducting with a cyclotron. A cyclotron is a particle accelerator, or atom smasher. In a cyclotron, small particles, such as protons, are made to travel at high speeds. The particles collide with atoms, causing the atoms to break apart into other elements.

Segrè's team suggested the name astatine for element 85 because there are no stable isotopes for the element. In Greek, the word for "unstable" is *astatos.*

Physical and Chemical Properties

The properties of astatine are not well known. The element breaks down too fast to allow experiments that take more than a few hours. Much of what is known about astatine comes from experiments conducted at the Argonne National Laboratory, outside Chicago, Illinois, and the Brookhaven

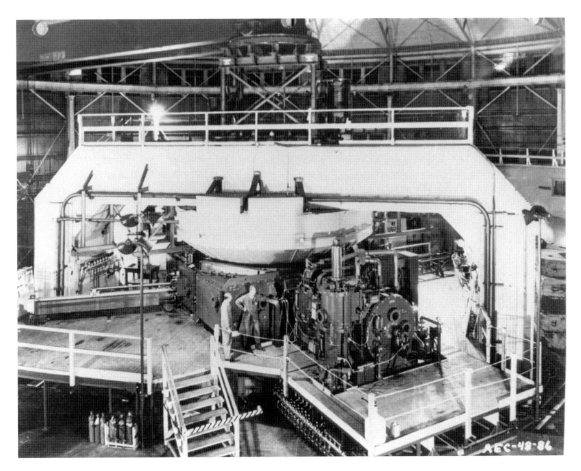

Overhead view of a cyclotron chamber. LIBRARY OF CONGRESS.

National Laboratory, in Upton, New York. Those experiments show that astatine is chemically similar to the other halogens above it in Group 17 of the periodic table. As chemists would expect, it acts more like a metal than iodine, the element just above it in the table. One of the few properties that have been determined for astatine is its melting point, found to be 576°F (302°C). Its boiling point is estimated to be about 639°F (337°C).

Occurrence in Nature

Astatine is produced in Earth's crust when the radioactive elements **uranium** and **thorium** decay. It can be made artificially only with great difficulty. By one estimate, no more than a millionth of a gram of astatine has ever been produced in the lab.

Isotopes

All 43 of astatine's isotopes are radioactive. That means they break down spontaneously and are transformed into other elements. Isotopes are two or more forms of an element. Isotopes differ from each other according to their mass number. The number written to the right of the element's name is the mass number. The mass number represents the number of protons plus neutrons in the nucleus of an atom of the element. The number of protons determines the element, but the number of neutrons in the atom of any one element can vary. Each variation is an isotope.

The isotopes with the longest half life are astatine-209, astatine-210, and astatine-211. The numbers after the names here are the atomic weights of the isotopes. These isotopes have half lives of 5.4 to 8.1 hours. The half life of a radioactive isotope is the time it takes for half of a sample of the element to break down. In 5.4 hours, only half of a sample of astatine-209 will still be astatine-209. Another 5.4 hours later, only 25 percent of it will remain.

Extraction

Astatine does not occur naturally.

Uses

Astatine is far too rare to have any uses. Some research suggests a possible medical use, however. Astatine is similar to the elements above it in Group 17 (VIIA) of the periodic table, especially iodine. One property of iodine is that it tends to collect in the thyroid gland. The thyroid is a gland at the base of the neck that controls many body functions.

Some researchers think that astatine will behave like iodine. If so, it could be used to treat certain diseases of the thyroid, such as thyroid cancer. When swallowed, the astatine would go to the thyroid. There, the radiation it gives off would kill cancer cells in the gland.

Compounds

There are no known commercial uses for astatine compounds.

Health Effects

As a radioactive element, astatine would pose a serious health hazard. However, because it can be produced only artificially—and with great difficulty at that—hardly anyone would ever be exposed to it.

Barium

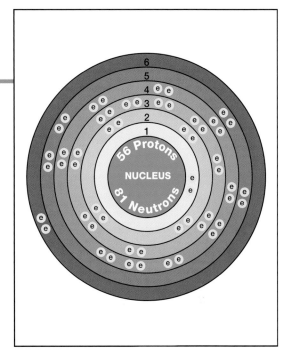

Overview

Barium was first isolated in 1808 by English chemist Sir Humphry Davy (1778–1829). In 1807 and 1808, Davy also discovered five other new elements: **sodium**, **potassium**, **strontium**, **calcium**, and **magnesium**. All of these elements had been recognized much earlier as new substances, but Davy was the first to prepare them in pure form.

Barium had first been identified as a new material in 1774 by Swedish chemist Carl Wilhelm Scheele (1742–1886). The form with which Scheele worked, however, was a compound of barium, barium sulfate ($BaSO_4$). Barium sulfate is, in fact, the most common naturally occurring ore of barium. It is generally known as barite or barytes.

Barium is a member of the alkaline earth metals. The alkaline earth metals make up Group 2 (IIA) of the periodic table. The other elements in this group are **beryllium**, magnesium, calcium, strontium, and **radium**. These elements tend to be relatively active chemically and form a number of important and useful compounds. They also tend to occur abundantly in Earth's crust in a number of familiar minerals such as aragonite, calcite, chalk, limestone, marble, travertine, magnesite, and dolomite. Alkaline earth compounds are widely used as building materials.

Key Facts

Symbol: Ba

Atomic Number: 56

Atomic Mass: 137.327

Family: Group 2 (IIA); alkaline earth metal

Pronunciation: BARE-ee-um

WORDS TO KNOW

Alkaline earth metal: An element found in Group 2 (IIA) of the periodic table.

Halogen: One of the elements in Group 17 (VIIA) of the periodic table.

Isotopes: Two or more forms of an element that differ from each other according to their mass number.

Malleable: Capable of being hammered into thin sheets.

Periodic table: A chart that shows how the chemical elements are related to each other.

Radiation: Energy transmitted in the form of electromagnetic waves or subatomic particles.

Radioactive isotope: An isotope that breaks apart and gives off some form of radiation.

Toxic: Poisonous.

Barium itself tends to have relatively few commercial uses. However, its compounds have a wide variety of applications in industry and medicine. Barium sulfate is used in X-ray studies of the gastrointestinal (GI) system. The GI system includes the stomach, intestines, and associated organs.

Discovery and Naming

The first mention of barium compounds goes back to the early 17th century. Early records mention a "Bologna stone," named for the city of Bologna, Italy. The Bologna stone glowed in the dark.

For more than 100 years, researchers labored without being able to identify the elements in the stone. In 1774, Scheele announced the presence of a new element in the Bologna stone. Today, scientists know that the stone was a form of barite. Five years later, Scheele demonstrated that barite was also present in heavy spar. This dense transparent mineral closely resembles ordinary spar, a compound of calcium.

Physical Properties

Pure barium is a pale yellow, somewhat shiny, somewhat malleable metal. Malleable means capable of being hammered into thin sheets. It has a melting point of about 1,300°F (700°C) and a boiling point of about 2,700°F (1,500°C). Its density is 3.6 grams per cubic centimeter.

When heated, barium compounds give off a pale yellow-green flame. This property is used as a test for barium.

Chemical Properties

Barium is an active metal. It combines easily with **oxygen**, the halogens, and other non-metals. The halogens are Group 17 (VIIA) of the periodic table and include **fluorine**, **chlorine**, **bromine**, **iodine**, and **astatine**. Barium also reacts with water and with most acids. It is so reactive that it must be stored under kerosene or some similar petroleum-based liquid to prevent it from reacting with oxygen and moisture in the air. Of the alkaline earth family, only radium is more reactive.

Occurrence in Nature

Barium is the 14th most abundant element in Earth's crust. Its abundance is estimated to be about 0.05 percent.

The most common sources of barium are barite and witherite. Witherite is an ore containing barium carbonate ($BaCO_3$). In 2008, the world's major sources of barite were China, India, the United States, Morocco, Iran, Mexico, and Turkey. Most of the barite processed in the United States came from Nevada and Georgia.

A barium X ray shows a patient with diverticulosis (an intestinal disorder).
PHOTOGRAPH BY MICHAEL ENGLISH. CUSTOM MEDICAL STOCK PHOTO.

Isotopes

There are seven naturally occurring isotopes of barium: barium-130, barium-132, barium-134, barium-135, barium-136, barium-137, and barium-138. Isotopes are two or more forms of an element. Isotopes differ from each other according to their mass number. The number written to the right of the element's name is the mass number. The mass number represents the number of protons plus neutrons in the nucleus of an atom of the element. The number of protons determines the element, but the number of neutrons in the atom of any one element can vary. Each variation is an isotope.

Thirty-nine radioactive isotopes of barium are known also. A radioactive isotope is one that breaks apart and gives off some form of radiation. Radioactive isotopes are produced when very small particles are fired at atoms. These particles stick in the atoms and make them radioactive.

None of the isotopes of barium has any practical commercial application.

Extraction

Pure barium is produced by reacting barium oxide (BaO) with **aluminum** or **silicon**:

$$3BaO + 2Al \rightarrow 3Ba + Al_2O_3$$

Uses

Barium metal has relatively few uses because it is so active. This activity makes it an excellent "getter" or "scavenger" when removing unwanted oxygen from sealed glass containers. (Oxygen can interfere with the operation of a vacuum tube, for example.) By adding a small amount of barium to the tube, the free oxygen inside will be "soaked up." The oxygen reacts with the barium to form barium oxide.

Compounds

Compounds of barium, especially barite ($BaSO_4$), are critical to the petroleum industry. Barite is used as a weighting agent in drilling new oil wells. A weighting agent is a material that adds body to petroleum.

Drilling for oil used to produce huge gushers. A gusher is when oil sprays out of the well into the air. Gushers are undesirable, because they waste oil and can burn for months if ignited.

Gushers are caused by the pressure of oil rushing into a newly drilled hole in the ground. This pressure pushes the oil upward much too rapidly. Barite is added to the hole as it is drilled. There, it tends to mix with oil in the ground and form a very dense mixture that comes out much more slowly and under control. Nearly 95 percent of the barite used in the United States was used by the petroleum industry in 2008.

Use in Medicine Perhaps the best known use of barium compounds is in medicine. Doctors often want to know what is happening inside a patient's body. One way to find out, of course, is through surgery. But surgery is a drastic procedure. It can cause new problems for the patient.

As a result, medical researchers have developed procedures that are less extreme. One such method is called radiography.

Radiography is a technique in which X rays are passed through the body. X rays are high energy light waves. They can pass through skin and tissue, but are absorbed by bones. So X rays are a good way of finding out if a bone is broken, for example.

Any type of light appearing on film from an X ray produces a black area, or exposure. The X rays pass through soft issues, exposing the film. Bones look greyish white on the film, depending on how much energy gets through.

Radiography can also be used for studying parts of the body where bones are *not* involved. For example, a doctor might want to study a person's stomach. Since there are no bones in the stomach, some other method must be used to see inside the stomach.

Barium sulfate is often used in such cases. Barium sulfate has some of the same properties as bony material. Therefore, since X rays will not pass through barium sulfate, this compound can be used to examine certain soft tissues.

Radiography using barium sulfate is called a barium swallow or a barium enema. Barium sulfate is mixed with water into a slurry (mixture) that looks and tastes like ground-up chalk. The patient swallows the dense mixture. A doctor or nurse then holds a fluoroscope over the patient's abdomen. The fluoroscope emits X rays that show up on a television screen.

The barium sulfate-water mixture slowly travels down the patient's throat, into the stomach, through the intestines, and out through the bowels. As the barium sulfate coats the lining of the digestive tract, a doctor can see if anything is wrong.

How can a toxic compound like barium sulfate be used for this procedure? Barium sulfate does not dissolve in water. So it cannot enter the bloodstream. If it cannot get into the blood, it has no toxic effects. The barium sulfate is eliminated through the bowels a few hours after the procedure.

Other Compound Uses Other uses of barite and other barium compounds include:

- barium sulfate (barite): used to add body to or as a coating for paper products; as a white coloring agent in paints, inks, plastics,

Since barium sulfate does not dissolve in liquids, it is often used to analyze the stomach. A patient swallows a water-barium sulfate mixture, and an X ray shows the path of the barium sulfate, highlighting any abnormalities. © LUIS DE LA MAZA/PHOTOTAKE NYC.

and textiles; in the manufacture of rubber products; in the production of batteries; in medical applications

- barium carbonate ($BaCO_3$): used in the production of chlorine and sodium hydroxide; as rat poison; in special types of glass
- barium oxide (BaO): used to remove water from solvents; in the petroleum industry
- barium nitrate ($Ba(NO_3)_2$): used in fireworks; as rat poison; in special ceramic glazes.

Health Effects

Barium and all of its compounds are very toxic.

Berkelium

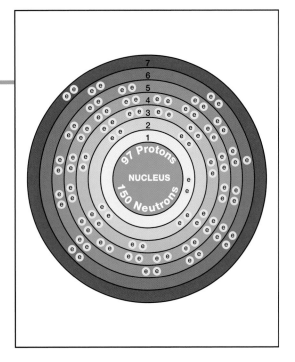

Overview

In the period between 1940 and 1961, 11 transuranium elements were discovered by researchers from the University of California at Berkeley (UCB). The term transuranium element refers to elements beyond **uranium** (atomic numbers greater than 92) in the periodic table. The periodic table is a chart that shows how chemical elements are related to each other. All transuranium elements are unstable or radioactive. Radioactive elements emit energy or particles as they decay into more stable atoms. One of these elements was berkelium.

Discovery and Naming

In 1949, element number 97 was produced in a particle accelerator on the UCB campus. A particle accelerator is sometimes called an atom smasher. It is used to speed up very small particles and atoms, which then collide with atoms of such elements as **gold**, **copper**, or **tin**. These atoms are called targets. When the particles strike target atoms precisely, the atom is converted into a new element.

The UCB researchers fired alpha particles—**helium** atoms without their electrons—at **americium** atoms in their particle accelerator. The result was a new element—number 97.

Key Facts

Symbol: Bk

Atomic Number: 97

Atomic Mass: [247]

Family: Actinoid; transuranium element

Pronunciation: BER-klee-um

WORDS TO KNOW

Actinoid family: Elements with atomic numbers 89 through 103.

Half life: The time it takes for half of a sample of a radioactive element to break down.

Isotopes: Two or more forms of an element that differ from each other according to their mass number.

Periodic table: A chart that shows how the chemical elements are related to each other.

Radioactive: Having a tendency to give off radiation.

Transuranium element: An element with an atomic number greater than 92.

helium + americium → new element (atomic number 97)

The new element was given the name berkelium by the UCB research team, in honor of the city of Berkeley, California, where the research was done.

Physical and Chemical Properties

Berkelium exists in such small amounts that very little is known about its properties. Scientists have found that it exists in two forms, known as the alpha form and the beta form. The melting points of these two forms of berkelium are 1920°F (1050°C) and 1810°F (986°C), respectively. The two forms have densities of 14.78 and 13.25 grams per cubic centimeter, about 14 times the density of water.

Occurrence in Nature

Berkelium does not occur in nature. It is made artificially.

Isotopes

All 17 known isotopes of berkelium are radioactive. The most stable is berkelium-247. Isotopes are two or more forms of an element. Isotopes differ from each other according to their mass number. The number written to the right of the element's name is the mass number. The mass number represents the number of protons plus neutrons in the nucleus of an atom of the element. The number of protons determines the element, but the number of neutrons in the atom of any one element can vary. Each variation is an isotope. A radioactive isotope is one that breaks apart and gives off some form of radiation.

Berkelium-247 has a half life of 1,380 years. The half life of a radioactive element is the time it takes for half of a sample of the element to break down. After 1,380 years, only half of a 10-gram sample (5 grams) of berkelium-247 would be left. The other half would have changed into a different element. After another 1,380 years, half of the remaining berkelium-247 would have changed, leaving 2.5 grams behind.

Extraction

Berkelium does not occur in nature. Therefore, it is not extracted.

Uses

Berkelium has no commercial uses.

Compounds

No compounds of any practical importance have been prepared.

Dr. Glenn T. Seaborg helped synthesize berkelium at UCB.
LIBRARY OF CONGRESS.

Health Effects

The health effects of berkelium have not been studied in detail. Since it is radioactive, scientists assume that it is harmful to human health.

Beryllium

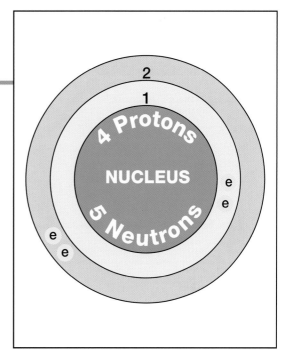

Overview

Beryllium is the lightest member of the alkaline earth metals family. These metals make up Group 2 (IIA) of the periodic table. They include beryllium, **magnesium**, **calcium**, **strontium**, **barium**, and **radium**. Elements in the same column of the periodic table have similar chemical properties. The periodic table is a chart that shows how the chemical elements are related to each other.

Beryllium was discovered by French chemist Louis-Nicolas Vauquelin (1763–1829) in 1798. Vauquelin suggested the name glucinium, meaning "sweet tasting," for the element because the element and some of its compounds have a sweet taste. The name beryllium was adopted officially in 1957.

Beryllium-**copper** alloys account for much of all the beryllium produced. An alloy is made by melting and mixing two or more metals. The mixture has properties different from those of the individual metals.

Discovery and Naming

A common compound of beryllium, beryl, was known in ancient Egypt, but nothing was known about the chemical composition of this mineral until the end of the 18th century. In 1797, French mineralogist

Key Facts

Symbol: Be

Atomic Number: 4

Atomic Mass: 9.012182

Family: Group 2 (IIA); alkaline earth metal

Pronunciation: buh-RIL-lee-um

WORDS TO KNOW

Alkaline earth metal: An element found in Group 2 (IIA) of the periodic table.

Alloy: A mixture of two or more metals with properties different from those of the individual metals.

Isotopes: Two or more forms of an element that differ from each other according to their mass number.

Laser: A device for making very intense light of one very specific color that is intensified many times over.

Periodic table: A chart that shows how chemical elements are related to each other.

Radioactive isotope: An isotope that breaks apart and gives off some form of radiation.

Toxic: Poisonous.

René-Just Haüy (1743–1822) completed studies on beryl and emerald. Emerald is a naturally occurring green gemstone. Haüy was convinced that these two minerals were nearly identical. He asked a friend, Vauquelin, to determine the chemical composition of the two minerals.

When Vauquelin performed his chemical analysis, he found a new material that had been overlooked because it is so much like **aluminum**. His data proved that the material was *not* aluminum. He suggested calling the new element glucinium. Scientists referred to the element by two different names, beryllium and glucinium, for 160 years. The name beryllium comes from the mineral, beryl, in which it was first discovered.

Physical Properties

Beryllium is a hard, brittle metal with a grayish-white surface. It is the least dense (lightest) metal that can be used in construction. Its melting point is 2,349°F (1,287°C) and its boiling point is estimated to be about 4,500°F (2,500°C). Its density is 1.8 grams per cubic centimeter, about twice that of water. The metal has a high heat capacity (it can store heat) and heat conductivity (it can transfer heat efficiently).

Beryllium is transparent in X rays. X rays pass through the metal without being absorbed. For this reason, beryllium is sometimes used to make the windows for X-ray machines.

Chemical Properties

Beryllium reacts with acids and with water to form **hydrogen** gas. It reacts with **oxygen** in the air to form beryllium oxide (BeO). The beryllium oxide then forms a thin skin on the surface of the metal that prevents the metal from reacting further with oxygen.

Occurrence in Nature

Beryllium never occurs as a free element, only as a compound. The most common ore of beryllium is beryl. Beryl has the chemical formula $Be_3(Al_2(SiO_3))_6$.

In 2008, the major beryl producer in the world was the United States, followed by China and Mozambique. In the United States, beryl was produced by only one mine in Utah. Some plants also converted beryl into beryllium and its compounds.

Beryllium is relatively common in Earth's crust. Its abundance is estimated at 2 to 10 parts per million.

Isotopes

Only one naturally occurring isotope of beryllium exists: beryllium-9. Isotopes are two or more forms of an element. Isotopes differ from each other according to their mass number. The number written to the right of the element's name is the mass number. The mass number represents the number of protons plus neutrons in the nucleus of an atom of the element. The number of protons determines the element, but the number of neutrons in the atom of any one element can vary. Each variation is an isotope.

Eight radioactive isotopes of beryllium are known also. A radioactive isotope is one that breaks apart and gives off some form of radiation. Radioactive isotopes are produced when very small particles are fired at atoms. These particles stick in the atoms and make them radioactive.

None of the isotopes of beryllium has any commercial use.

Extraction

Beryllium ores are first converted to beryllium oxide (BeO) or beryllium hydroxide ($Be(OH)_2$). These compounds are then converted to beryllium chloride ($BeCl_2$) or beryllium fluoride (BeF_2). Finally, the pure metal is isolated by: (1) an electric current:

$$BeCl_2 \xrightarrow{\text{electric current}} Be + Cl_2$$

or, (2) reaction with magnesium metal at high temperature:

$$BeCl_2 + Mg \rightarrow MgCl_2 + Be$$

Uses

By far the greatest use of beryllium metal is in alloys. Beryllium alloys are popular because they are tough, stiff, and lighter than similar alloys. For example, an alloy of beryllium and aluminum called Beralcast was released in 1996. Beralcast is three times as stiff and 25 percent lighter than pure aluminum. It is used in helicopters and satellite guidance systems.

Other popular alloys of beryllium are those with copper metal. Copper-beryllium alloys contain about 2 percent beryllium. They conduct heat and electricity almost as well as pure copper but are stronger, harder, and more resistant to fatigue (wearing out) and corrosion (rusting). These alloys are used in circuit boards, radar, computers, home appliances, aerospace applications, automatic systems in factories, automobiles, aircraft landing systems, oil and gas drilling equipment, and heavy machinery.

Gemstones Beryllium is also associated with gemstones. A gemstone is a mineral that can be cut and polished for use in jewelry. Some typical gemstones are jade, sapphire, diamond, ruby, amethyst, emerald, spinel, moonstone, topaz, aquamarine, opal, and turquoise. Gemstones are often used as birthstones, which honor the month in which a person is born. (For instance, the birthstone for April is a diamond.)

Gemstones are valued for their beautiful colors and crystal forms. Light reflects off them in brilliant patterns. The crystal forms are the result of very exact arrangements of atoms in the gemstone. Its perfection contributes to its beauty and its monetary value.

But gemstone color is due to very small impurities in the mineral. For example, the mineral known as corundum is colorless when pure. But a very small amount of **chromium** produces a bright red color. The corundum is now a ruby. A touch of **iron** or **titanium** produces shades of yellow, green, purple, pink, or blue that turn it into a sapphire.

Two gemstones are made primarily of beryl. They are emeralds and aquamarines. In emeralds, traces of chromium produce a brilliant green color. In aquamarines, iron is the impurity. It gives the beryl a beautiful blue color.

Compounds

Some of the beryllium used in the United States is in the form of beryllium oxide (BeO). Beryllium oxide is a white powder that can be made into many different shapes. It is desirable as an electrical insulator because

it conducts heat well, but an electrical current poorly. It is used in high-speed computers, auto ignition systems, lasers, microwave ovens, and systems designed to hide from radar signals.

Health Effects

Beryllium is a very toxic metal. It is especially dangerous in powder form. The effects of inhaling beryllium powder can be acute or chronic. Acute effects are those that occur very quickly as the result of large exposures. Chronic effects are those that occur over very long periods of time as the result of much smaller exposures. Acute effects of inhaling beryllium powder include pneumonia-like symptoms that can result in death in a short time. Chronic effects include diseases of the respiratory system (throat and lungs), such as bronchitis and lung cancer.

These effects can be avoided fairly easy. Workers can wear masks over their faces to filter out beryllium particles. Filtering devices in factories where beryllium is used also prevent beryllium from getting into the air.

Bismuth

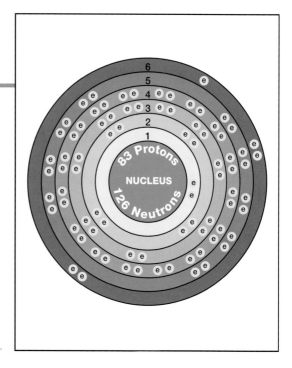

Overview

Early chemists had difficulty separating similar elements from each other. Elements with similar properties can be told apart only with tests not available before the 18th century.

Chemists also believed that metals grew in the earth, in much the same way that plants grow. Unattractive metals, like **lead**, were thought to be young or immature metals. More attractive metals, like **tin**, were thought to be partially grown. The most mature metals were **silver** and **gold**. This made identification very difficult. Were chemists looking at "older lead" or a "younger tin?"

Bismuth is one of the elements often confused with other elements. Old manuscripts show that bismuth was often confused with lead, tin, **antimony**, or even silver.

Bismuth was used in early alloys. An alloy is made by melting and mixing two or more metals. The mixture has properties different from those of the individual metals. The first printing presses (dating back to the 1450s) used type made of bismuth alloys.

Key Facts

Symbol: Bi

Atomic Number: 83

Atomic Mass: 208.98040

Family: Group 15 (VA); nitrogen

Pronunciation: BIZ-muth

WORDS TO KNOW

Alchemy: A kind of pre-science that existed from about 500 BCE to about the end of the 16th century.

Alloy: A mixture of two or more metals with properties different from those of the individual metals.

Isotopes: Two or more forms of an element that differ from each other according to their mass number.

Periodic table: A chart that shows how the chemical elements are related to each other.

Radiation: Energy transmitted in the form of electromagnetic waves or subatomic particles.

Radioactive isotope: An isotope that breaks apart and gives off some form of radiation.

Toxic: Poisonous.

Discovery and Naming

As with **arsenic** and antimony, it is difficult to say who exactly discovered bismuth. The name bismuth was probably taken from two German words, *weisse masse,* meaning "white mass." The phrase describes how the element appears in nature. Later the name was shortened to *wismuth,* and then to *bisemutum,* before bismuth came into common use.

In 1753, French scholar Claude-Françoise Geoffrey wrote a book summarizing everything that was known about bismuth at the time.

Physical Properties

Bismuth is a soft, silvery metal with a bright, shiny surface and a yellowish or pinkish tinge. The metal breaks easily and cannot be fabricated (worked with) at room temperature. Its melting point is 520°F (271°C) and its boiling point is 2,840°F (1,560°C). Its density is 9.78 grams per cubic centimeter.

Bismuth expands as it solidifies (changes from a liquid to a solid). Most materials contract (have a smaller volume) as they solidify. Few elements behave like bismuth.

This property makes bismuth useful for producing type metal. An alloy of bismuth is melted and poured into molds that have the shape of letters and numbers. As the type cools, it solidifies and expands to fill all the corners of the mold. The type formed is clear, crisp, and easy to read. Computer typesetting, however, has largely replaced bismuth metal typesetting.

Chemical Properties

Bismuth combines slowly with **oxygen** at room temperature. Bismuth oxide (Bi_2O_3) gives the metal its pinkish or yellowish tinge. At higher temperatures, the metal burns to form bismuth oxide. Bismuth also reacts with most acids.

Occurrence in Nature

The abundance of bismuth in Earth's crust is estimated to be about 0.2 parts per million, making it a relatively rare element. This puts it in the bottom quarter of the elements according to their abundance in the earth.

Bismuth is seldom found in its elemental state (as a pure metal) in the earth. Its compounds are generally found along with ores of other metals, such as lead, silver, gold, and **cobalt**. The most important mineral of bismuth is bismuthinite, also known as bismuth glance (Bi_2S_3).

As of 2008, the largest producers of bismuth in the world were China, Mexico, Peru, and Canada. Bismuth is produced in the United States as a by-product of lead refining.

Isotopes

There is only one naturally occurring isotope of bismuth: bismuth-209. Isotopes are two or more forms of an element. Isotopes differ from each other according to their mass number. The number written to the right of the element's name is the mass number. The mass number represents the number of protons plus neutrons in the nucleus of an atom of the element. The number of protons determines the element, but the number of neutrons in the atom of any one element can vary. Each variation is an isotope.

Fifty-nine radioactive isotopes of bismuth are known also. A radioactive isotope is one that breaks apart and gives off some form of radiation. Radioactive isotopes are produced when very small particles are fired at atoms. These particles stick in the atoms and make them radioactive.

None of the radioactive isotopes of bismuth have any commercial applications.

Extraction

Bismuth metal is usually separated from ores of other metals by the Betterton-Kroll process. **Calcium** or **magnesium** is added to the molten (melted) ore, forming an alloy with bismuth. Later, the bismuth can be separated from the calcium or magnesium to make the pure metal.

In 2008, the price of bismuth ranged from approximately $9.75 to $13.25 per pound.

Uses

The primary use of bismuth metal is in making alloys. Many bismuth alloys have low melting points. The metal itself melts at 520°F (271°C), but some bismuth alloys melt at temperatures as low as 160°F (70°C). This temperature is below the boiling point of water. These alloys are used in fire sprinkler systems, fuel tank safety plugs, solders, and other applications.

Interest in using bismuth as a substitute for lead in alloys has increased. Lead is toxic to humans and other animals so scientists are trying to find ways to replace lead in most applications. For example, an alloy containing 97 percent bismuth and 3 percent tin is popular as shot used in waterfowl hunting. Bismuth is also being used in place of lead in plumbing applications and in coloring paints, ceramics, and glazes.

Compounds

About a third of bismuth produced in the United States is made into drugs, pharmaceuticals, and other chemicals. The most widely used compound is bismuth subsalicylate ($Bi(C_7H_5O_3)_3$), the active ingredient in many over-the-counter stomach remedies. An over-the-counter drug is one that can be sold without a prescription.

Other compounds used in medicine include bismuth ammonium citrate $(Bi(NH_4)_3(C_6H_5O_7)_2)$, bismuth citrate $(BiC_6H_5O_7)_2)$, bismuth subgallate $(Bi(OH)_2OOCC_6H_2(OH)_3)$, and bismuth tannate. These compounds are used to treat a large variety of problems, including burns, stomach ulcers, and intestinal disorders, and in veterinary applications.

Bismuth compounds are also widely used in various cosmetics. Bismuth oxychloride $(BiOCl)$ is a lustrous white powder added to face powder. Bismuth subcarbonate $[(BiO)_2CO_3]$ and bismuth subnitrate $[4BiNO_3(OH)_2 \cdot BiO(OH)]$ are also white powders used to give a pearl-like luster to lipstick, eyeshadow, and other cosmetics.

Health Effects

Bismuth and its compounds are not thought to be health hazards. In fact, bismuth compounds are used in medications. A relatively new bismuth compound is used to treat ulcers, a common stomach problem.

Boron

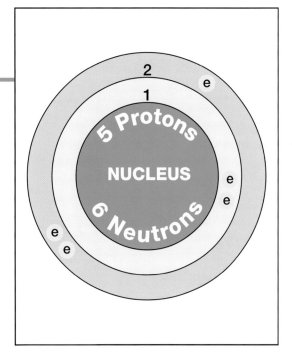

Overview

Boron is the first element in Group 13 (IIIA) of the periodic table. The periodic table is a chart that shows how the chemical elements are related to each other. The elements in this group are usually referred to as the aluminum family.

Boron is quite different from other members of the family. One difference is that boron is not a metal. All other members of the family (**aluminum**, **gallium**, **indium**, and **thallium**) are metals.

Compounds of boron have been used for centuries. Borax, a boron compound, has long been used to make glass and glazes. The element itself was not identified until 1808.

The most important use of boron is still in glass manufacture. Agricultural products, fire retardants, and soaps and detergents are all made of boron compounds.

Discovery and Naming

The first mention of boron compounds is found in a book by Persian alchemist Rhazes (c. 865–c. 925). Alchemists studied the nature of

Key Facts

Symbol: B

Atomic Number: 5

Atomic Mass: 10.811

Family: Group 13 (IIIA)

Pronunciation: BOR-on

WORDS TO KNOW

Abrasive: A powdery material used to grind or polish other materials.

Alloy: A mixture of two or more metals with properties different from those of the individual metals.

Alpha radiation: A form of radiation that consists of very fast moving alpha particles and helium atoms without their electrons.

Isotopes: Two or more forms of an element that differ from each other according to their mass number.

Periodic table: A chart that shows how the chemical elements are related to each other.

Radioactive isotope: An isotope that breaks apart and gives off some form of radiation.

Refractory: A material that can withstand very high temperatures and reflect heat away from itself.

matter before modern chemistry was born. Rhazes classified minerals into six classes, one of which was the *boraces,* which included borax.

Borax was widely used by crafts people. It reduces the melting point of materials used to make glass. It was also used to melt the ores of metals and to isolate the metals from those ores.

In 1808, English chemist Humphry Davy (1778–1829) had just learned how to isolate the most active metals, such as **sodium** and **potassium**. He was also working on a method to remove boron from its compounds.

News of Davy's success had traveled to France, where emperor Napoleon Bonaparte (1769–1821) grew concerned about the scientific reputation of his country. He ordered larger and better equipment built for his scientists. He wanted them to surpass Davy in his work on metals. This equipment was designed especially for two French chemists, Louis Jacques Thênard (1777–1857) and Joseph Louis Gay-Lussac (1778–1850).

Thênard and Gay-Lussac found a new way to separate boron from its compounds. They heated boracic acid (also known as boric acid, H_3BO_3) with potassium metal to produce impure boron. Thênard and Gay-Lussac were given credit for discovering the new element. In 1892, French chemist Henri Moissan (1852–1907) produced boron that was 98 percent pure.

The names borax and boracic acid probably originated as far back as the time of Rhazes as *buraq* (in Arabic) or *burah* (in Persian).

Physical Properties

One of the unusual properties of boron is the many physical forms, called allotropes, in which it occurs. Allotropes are forms of an element with different physical and chemical properties. One form of boron consists of clear red crystals with a density of 2.46 grams per cubic centimeter. A second form consists of black crystals with a metallic appearance and a density of 2.31 grams per cubic centimeter. Boron can also occur as a brown powder with no crystalline structure. The density of this powder is 2.350 grams per cubic centimeter.

All forms of boron have very high melting points, from 4,000 to 4,200°F (2,200 to 2,300°C).

One property of special importance is boron's ability to absorb neutrons. Neutrons are subatomic particles with no charge that occur in the nucleus of nearly all atoms. Boron atoms are able to absorb a large number of neutrons. This makes boron useful in the control rods of nuclear reactors.

A nuclear reactor is a device for generating energy from nuclear fission reactions. Nuclear fission is the process in which large atoms are split, releasing large amounts of energy and smaller atoms. In a nuclear reactor, it is essential that just the right number of neutrons are present. Too many neutrons can cause a fission reaction to get out of control. Too few neutrons and a fission reaction stops.

Control rods are long tubes packed with boron (or some other element). The rods can be raised and lowered in the reactor. As the rods are lowered into the core, the boron absorbs neutrons, slowing the reaction.

Chemical Properties

Boron combines with **oxygen** in the air to form boron trioxide (B_2O_3). Boron trioxide forms a thin film on the element's surface that prevents further reaction with oxygen.

Boron is not soluble in water. It normally does not react with acids. In powder form, it reacts with hot nitric acid (HNO_3) and hot sulfuric acid (H_2SO_4). It also dissolves in molten (melted) metals.

Occurrence in Nature

The abundance of boron in Earth's crust is estimated to be about 10 parts per million. That places it in about the middle among the elements in terms of their abundance in the earth.

Boron never occurs as a free element but always as a compound. The most common minerals of boron are borax, or sodium borate ($Na_2B_4O_7$); kernite (another form of sodium borate); colemanite, or **calcium** borate ($Ca_2B_6O_{11}$); and ulexite, or sodium calcium borate ($NaCaB_5O_9$). These minerals usually occur as white crystalline deposits in desert areas. As of 2008, Turkey was the largest producer of boron ore. Other major producers of boron materials are Argentina, Chile, Russia, China, Bolivia, and Kazakhstan. Production statistics for the United States were not released in order to protect trade secrets.

Isotopes

Two naturally occurring isotopes of boron exist: boron-10 and boron-11. Isotopes are two or more forms of an element. Isotopes differ from each other according to their mass number. The number written to the right of the element's name is the mass number. The mass number represents the number of protons plus neutrons in the nucleus of an atom of the element. The number of protons determines the element, but the number of neutrons in the atom of any one element can vary. Each variation is an isotope. Boron-10 is the isotope with high neutron-absorbing tendencies described earlier under "Physical Properties."

Chemical Elements, 2nd Edition

Nine radioactive isotopes of boron are known also. A radioactive isotope is one that breaks apart and gives off some form of radiation. Radioactive isotopes are produced when very small particles are fired at atoms. These particles stick in the atoms and make them radioactive.

None of the radioactive isotopes of boron have any important commercial uses.

Extraction

Boron is still produced by a method similar to that used by Thênard and Gay-Lussac. Boric oxide is heated with powdered **magnesium** or aluminum:

$$2Al + B_2O_3 \ —heat\rightarrow \ Al_2O_3 + 2B$$

The element can also be obtained by passing an electric current through molten (melted) boron trichloride:

$$2BCl_3 \ —electrical\ current\rightarrow \ 2B + 3Cl_2$$

Uses

Boron is used to make certain types of alloys. An alloy is made by melting and mixing two or more metals. The mixture has properties different

Experimental Cancer Treatment

Boron is also associated with cancer treatment, an area dominated by radiation. Radiation can kill living cells. Light, X rays, radio waves, and microwaves are all forms of radiant energy. These forms of radiation differ from each other in the amount of energy they carry with them. X rays carry a great deal of energy; light waves, less energy; and radio waves, very little energy.

The bad news about high-energy radiation is that it can kill healthy cells. A person exposed to high levels of X rays will become ill and may die. Because the X rays kill so many cells, the person's body cannot survive. Essential body functions stop, and death occurs.

The good news is that high-energy radiation can be used to kill cancer cells. Cancer cells are abnormal cells that reproduce faster than normal cells. The rapidly dividing cells form tumors, crowd organs, and shut down some organ functions. Radiation is one way to kill cancer cells.

The problem lies in killing only the cancer cells. The radiation has to be "targeted" at the cancer (bad) cells, and not the healthy (good) cells. Scientists think that using boron may be one way of achieving this goal. The experimental

procedure called boron neutron capture therapy (BNCT) is one method for targeting cancer cells.

With BNCT, a person with cancer receives an injection of boron-10. The boron tends to go directly to cancer cells. Scientists currently do not know why boron favors cancer cells. But it does.

The patient's body is then bombarded with neutrons that pass through tissue without harming healthy cells. They then collide with boron atoms. The boron is converted into **lithium** atoms, alpha particles, and gamma rays. An alpha particle is a **helium** atom without electrons. Gamma radiation is very high-energy radiation that can kill cells.

The lithium atoms and alpha particles travel only a short distance. They do not leave the cancer cell but have enough energy to kill the cell. Since they do not leave the cell, they pose no threat to healthy cells nearby.

BNCT is not fully developed. But it holds great promise as a cancer treatment. It is currently being used primarily to treat two forms of cancer: glioblastoma, a very dangerous form of cancer that can not be treated by other means, and malignant melanoma, a very serious form of skin cancer.

from those of the individual metals. The most important of these alloys commercially are used to make some of the strongest magnets known. The rare earth magnets, for example, are made from boron, **iron**, and **neodymium**. These magnets are used for microphones, magnetic switches, loudspeakers, headphones, particle accelerators, and many other technical applications.

The use of boron in nuclear power plants was described earlier under "Physical Properties."

Compounds

The most important boron compound is sodium borate ($Na_2B_4O_7$), used in the manufacture of borosilicate glass, glass fiber insulation, and textile glass fiber. The addition of sodium borate to glass makes it easier to work while it is molten. The final glass is not attacked by acids or water, is very strong, and resists thermal shock. Resistance to thermal shock means the glass can be heated and cooled very quickly without breaking. The Pyrex glass used in kitchenware and chemistry laboratories is a form of borosilicate glass. High quality optical glass, like that used in telescopes, is also made from borosilicate glass.

Glass fiber insulation is made by forcing borosilicate glass through narrow openings. The glass comes out as a thin fiber and is then spun into insulation. These fibers trap air. Since neither the borosilicate fibers nor air is a good conductor of heat, it makes excellent insulation. Much of the insulation used in private homes, office buildings, and other structures is made of borosilicate fibers.

Fibers made from borosilicate glass are also used in making cloth. Borosilicate fibers are blended with other synthetic fibers to make durable fabric for automobile seat covers and other long-wear applications.

Boron also forms important compounds with two other elements, **carbon** and **nitrogen**. Boron carbide (B_4C) and boron nitride (BN) are important compounds because of their hardness. In fact, boron nitride may be the hardest substance known. Both compounds have very high melting points: 4,262°F (2,350°C) for boron carbide and more than 5,432°F (3,000°C) for boron nitride.

These properties make boron carbide and boron nitride useful as abrasives and refractories. An abrasive is a powdery material used to grind or polish other materials. A refractory material is one that can withstand very high temperatures by reflecting heat. Refractory materials line ovens to maintain high temperatures.

Boron carbide and boron nitride are used in high-speed tools, military aircraft and spacecraft, heat shields, and specialized heat-resistant fibers. They also are found in face powders, cream make-ups, and lipsticks.

Small amounts of boron compounds are also used to control the growth of weeds in agriculture, and as insecticides, fertilizers, and flame retardants. A flame retardant is a material that prevents another material from catching fire and burning with an open flame.

Health Effects

The role of boron in human health is not well understood. There is growing evidence that very small amounts of boron may be required to maintain healthy bones, especially in women. Studies suggest that a lack of boron may lead to arthritis and other disorders of the skeleton. Boron may also be necessary for healthy brain functions, such as memory and hand-eye coordination.

No specific recommendations have been made by health authorities. But some experts believe boron should be included in the daily diet. Most people get boron from fruits, green vegetables, nuts, and beans in their normal diet.

Boron is also an essential trace mineral in plants. A trace mineral is an element needed in minute amounts for the good health of an organism. Boron is critical to production of certain essential plant proteins and to help plants extract water from the soil. Low levels of boron show up as yellowing, blackening, twisting, or crumpling of leaves.

Few studies have been conducted on the harmful effects of exposure to boron and its compounds. Some evidence suggests that the element and its compounds may cause irritation of the eyes, skin, and respiratory system. No serious long-term problems, such as cancer, have yet been associated with exposure to boron or its compounds.

Bromine

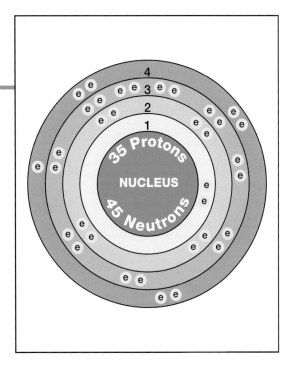

Overview

Bromine is a member of the halogen family. Halogens are the elements that make up Group 17 (VIIA) of the periodic table. The periodic table is a chart that shows how elements are related to one another. The word halogen means "salt-former." **Fluorine**, **chlorine**, bromine, **iodine**, and **astatine** form salts when chemically combined with a metal.

Bromine was discovered, at almost the same time in 1826, by two men, German chemist Carl Löwig (1803–1890) and French chemist Antoine-Jérôme Balard (1802–1876). Although Balard announced his discovery first, Löwig had simply not completed his studies of the element when Balard made his announcement.

The vast majority of all bromine produced comes from the United States, Israel, and China. In 2008, Israel produced about 182,000 short tons (165,000 metric tons) and China contributed 149,000 short tons (135,000 metric tons). Statistics for the United States were not released in order to protect trade secrets.

The largest single use of the element is in the manufacture of flame retardants. Flame retardants are chemicals added to materials to prevent burning or to keep them from burning out of control. Other major uses

Key Facts

Symbol: Br

Atomic Number: 35

Atomic Mass: 79.904

Family: Group 17 (VIIA); halogen

Pronunciation: BRO-meen

WORDS TO KNOW

Density: The mass of a substance per unit volume.

Halogen: One of the elements in Group 17 (VIIA) of the periodic table.

Isotopes: Two or more forms of an element that differ from each other according to their mass number.

Periodic table: A chart that shows how the chemical elements are related to each other.

Radioactive isotope: An isotope that breaks apart and gives off some form of radiation.

Toxic: Poisonous.

are in the manufacture of drilling fluids, pesticides, chemicals for the purification of water, photographic chemicals, and as an additive to rubber.

Discovery and Naming

Compounds of bromine had been known for hundreds of years before the element was discovered. One of the most famous of these compounds was Tyrian purple, also called royal purple. (Tyrian comes from the word Tyre, an ancient Phoenician city.) Only very rich people or royalty could afford to buy fabric dyed with Tyrian purple. It was obtained from a mollusk (shellfish) found on the shores of the Mediterranean Sea (a large body of water bordered by Europe, Asia, and Africa).

In 1825, Löwig enrolled at the University of Heidelberg in Germany to study chemistry. He continued an experiment he had begun at home in which he added chlorine to spring water. The addition of ether to that mixture produced a beautiful red color. Löwig suspected he had discovered a new substance. A professor encouraged him by suggesting he study the substance in more detail.

As these studies progressed, Balard published a report in a chemical journal that announced the discovery of the new element bromine. The element had all the properties of Löwig's new substance. The two chemists had made the discovery at nearly the same time. Balard, however, is credited as the discoverer of bromine, because scientists acknowledge the first person to *publish* his or her findings.

In Greek, the word *bromos* means "stench" (strong, offensive odor). Bromine lives up to the description. The odor is intense and highly irritating to the eyes and lungs.

Chemists found that bromine belonged in the halogen family. They knew that it had properties similar to other halogens and placed it below fluorine and chlorine in the periodic table.

Physical Properties

Only two liquid elements exist—bromine and **mercury**. At room temperature, bromine is a deep reddish-brown liquid. It evaporates easily, giving off strong fumes that irritate the throat and lungs. Bromine boils at 137.8°F (58.8°C), and its density is 3.1023 grams per cubic centimeter. Bromine freezes at 18.9°F (–7.3°C).

Bromine dissolves well in organic liquids—such as ether, alcohol, and carbon tetrachloride—but only slightly in water. Organic compounds contain the element **carbon**.

Chemical Properties

Bromine is a very reactive element. Although it is less reactive than fluorine or chlorine, it is more reactive than iodine. It reacts with many metals, sometimes very vigorously. For instance, with **potassium**, it reacts explosively. Bromine even combines with relatively unreactive metals, such as **platinum** and **palladium**.

Occurrence in Nature

Bromine is too reactive to exist as a free element in nature. Instead, it occurs in compounds, the most common of which are **sodium** bromide (NaBr) and potassium bromide (KBr). These compounds are found in seawater and underground salt beds. These salt beds were formed in regions where oceans once covered the land. When the oceans evaporated (dried up), salts were left behind—primarily sodium chloride (NaCl), potassium chloride (KCl), and sodium and potassium bromide. Later, movements of Earth's crust buried the salt deposits. Now they are buried miles underground. The salts are brought to the surface in much the same way that coal is mined.

Bromine is a moderately abundant element. Its abundance in Earth's crust is estimated to be about 1.6 to 2.4 parts per million. It is far more abundant in seawater where its occurrence is estimated to be about 65 parts per million.

In some regions, the abundance of bromine is even higher. For example, the Dead Sea (which borders Israel and Jordan), has a high level of dissolved salts. The abundance of bromine there is estimated to be 4,000 parts per million. The salinity, or salt content, is so high that nothing lives in the water. That fact explains how the Dead Sea got its name.

Isotopes

Two naturally existing isotopes of bromine exist: bromine-79 and bromine-81. Isotopes are two or more forms of an element. Isotopes differ from each other according to their mass number. The number written to the right of the element's name is the mass number. The mass number represents the number of protons plus neutrons in the nucleus of an atom of the element. The number of protons determines the element, but the number of neutrons in the atom of any one element can vary. Each variation is an isotope.

Thirty-four radioactive isotopes of bromine are known also. A radioactive isotope is one that breaks apart and gives off some form of radiation. Radioactive isotopes are produced when very small particles are fired at atoms. These particles stick in the atoms and make them radioactive.

No isotope of bromine has any important commercial use.

Extraction

The method used by Löwig and Balard to collect bromine continues to be used today. Chlorine is added to seawater containing sodium bromide or potassium bromide. Chlorine is more active than bromine and replaces bromine in the reaction:

$$Cl_2 + 2NaBr \rightarrow 2NaCl + Br_2$$

Uses and Compounds

The most important use of bromine today is in making flame retardant materials. Many materials used in making clothing, carpets, curtains, and drapes are flammable, and if a flame touches them, they burn very quickly. Chemists have learned how to make materials more resistant to fires by soaking them in a bromine compound. The compound coats the fibers of the material. The bromine compound can also be chemically incorporated into the material.

The bromine compounds used in flame retardants are often complicated. One such compound is called tris(dibromopropyl)phosphate $((Br_2C_3H_5O)_3PO)$. However, this compound has been found to be a carcinogen (cancer-causing substance). Its use, therefore, has been severely restricted.

Bromine is also used in drilling wells. Calcium bromide $(CaBr_2)$, sodium bromide $(NaBr)$, or zinc bromide $(ZnBr_2)$ is added to a well to increase the efficiency of the drilling process.

Bromine is also important in the manufacture of pesticides, chemicals used to kill pests. For many years, one of the most popular bromine-based pesticides was methyl bromide (CH_3Br). Methyl bromide is sprayed on the surface or injected directly into the ground to control pests.

The problem with using methyl bromide is that some of the chemical always evaporates into the air. It then escapes into Earth's stratosphere, where it damages the ozone layer. Ozone (O_3) gas filters out a portion of the ultraviolet (UV) radiation from the sun. UV radiation causes skin cancer, sunburn, and damage to plants and fragile organisms.

Bromine compounds increase the efficiency of the drilling process in oil rigs.
IMAGE COPYRIGHT 2009, BRAM VAN BROEKHOVEN. USED UNDER LICENSE FROM SHUTTERSTOCK.COM.

Because of the damage it causes to the environment, many countries have banned the production and/or use of methyl bromide as a pesticide. The U.S. ban on methyl bromide became effective on January 1, 2005. Some exceptions to the ban exist, permitting the pesticide's use for certain specialized purposes.

Ethylene dibromide ($C_2H_4Br_2$) is a bromine compound added to leaded gasoline. The lead in "leaded gasoline" is tetraethyl lead ($Pb(C_2H_5)_4$). It helps fuels burn more cleanly and keeps car engines from "knocking." "Knocking" is a repetitive metallic banging sound that occurs when there are ignition problems with a car's engine. "Knocking" reduces the efficiency of a car engine.

But leaded gasoline gives off free lead as it burns. Free lead is a very toxic element that causes damage to the nervous system. Ethylene dibromide is added to react with free lead and convert it to a safe compound.

Ethylene dibromide does not completely solve the problem. Some free lead still escapes into the atmosphere. Leaded gasoline has been banned in the United States for many years but is still used in some countries.

Bromine is also used as a disinfectant for swimming pools and water used in industrial cooling towers. It kills disease-causing bacteria as effectively as chlorine, which was once the most popular disinfectant for these purposes. But it adds less odor to water than chlorine and is less likely to irritate a person's skin, eyes, and nose. Today, many individuals and companies have switched from chlorine to bromine for disinfecting the water in swimming pools and cooling towers.

Health Effects

Bromine is toxic if inhaled or swallowed. It can damage the respiratory system and the digestive system, and can even cause death. It can also cause damage if spilled on the skin.

Cadmium

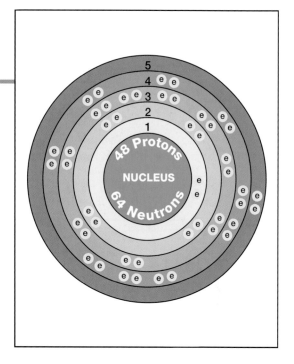

Overview

Cadmium is a transition metal. The transition metals are the elements found in Rows 4 through 7 between Groups 2 and 13 in the periodic table, a chart that shows how elements are related to each other. Cadmium was discovered by German chemist Friedrich Stromeyer (1776–1835) in 1817. It is found most commonly in ores of **zinc**.

Cadmium is a soft metal that is easily cut with a knife. It resembles zinc in many of its physical and chemical properties. However, it is much less abundant in Earth's crust than zinc.

Among the most important uses of cadmium in the United States is in the production of nicad (**nickel**-cadmium), or rechargeable, batteries. It is also used in pigments, coatings and plating, manufacture of plastic products, and alloys. An alloy is made by melting and mixing two or more metals. The mixture has properties different from those of the individual metals.

Caution must be used when handling cadmium and its compounds, as they are toxic to humans and animals. They present a threat to the environment because they are used for so many applications.

Key Facts

Symbol: Cd

Atomic Number: 48

Atomic Mass: 112.411

Family: Group 12 (IIB); transition metal

Pronunciation: CAD-mee-um

WORDS TO KNOW

Alloy: A mixture of two or more metals with properties different from those of the individual metals.

Isotopes: Two or more forms of an element that differ from each other according to their mass number.

Periodic table: A chart that shows how the chemical elements are related to each other.

Radioactive isotope: An isotope that breaks apart and gives off some form of radiation.

Toxic: Poisonous.

Transition metal: An element in Groups 3 through 12 of the periodic table.

Discovery and Naming

In addition to being a professor at Göttingen University, Stromeyer was a government official responsible for inspecting pharmacies in the state of Hanover, Germany. On one inspection trip, he found that many pharmacies were stocking a compound of zinc called zinc carbonate ($ZnCO_3$) instead of the usual zinc oxide (ZnO).

Stromeyer was told that the supplier had problems making zinc oxide from zinc carbonate and had offered the substitution. The normal process was to heat zinc carbonate to produce zinc oxide:

$$ZnCO_3 \longrightarrow heat \rightarrow ZnO + CO_2$$

The supplier explained that zinc carbonate turned yellow when heated. Normally, a yellow color meant that **iron** was present as an impurity. The supplier found no iron in his zinc carbonate, but it was still yellow. Pharmacies would not buy yellow zinc oxide, so the supplier sold white zinc carbonate instead.

Stromeyer analyzed the odd yellow zinc carbonate. What he discovered was a new element—cadmium. The cadmium caused the zinc carbonate to turn yellow when heated. The name cadmium comes from the ancient term for zinc oxide, *cadmia*. Zinc oxide is still available in pharmacies today. It is sold under the name of calamine lotion. Calamine lotion is a popular remedy for stopping the itch of sunburn or bug bites.

Physical Properties

Cadmium is a shiny metal with a bluish cast (shade) to it. It is very soft and can almost be scratched with a fingernail. Its melting point is

Cadmium samples. © RICH TREPTOW, NATIONAL AUDUBON SOCIETY COLLECTION/PHOTO RESEARCHERS, INC.

610°F (321°C) and its boiling point is 1,410°F (765°C). The density of cadmium is 8.65 grams per cubic centimeter.

An interesting property of cadmium is its effect in alloys. In combination with certain metals, it lowers the melting point. Some common low-melting-point alloys are Lichtenberg's metal, Abel's metal, Lipowitz' metal, Newton's metal, and Wood's metal. The Wood's metal alloy melts at 158°F (70°C), and is used in fire sprinkler systems as a plug. When the temperature rises above 158°F (70°C), the plug melts and falls out. This opens up the water line and activates the sprinkler. Out sprays the water!

Chemical Properties

Cadmium reacts slowly with **oxygen** in moist air at room temperatures, forming cadmium oxide:

$$2Cd + O_2 \rightarrow 2CdO$$

Cadmium does not react with water, although it reacts with most acids.

Occurrence in Nature

The abundance of cadmium in Earth's crust is estimated to be about 0.1 to 0.2 parts per million. It ranks in the lower 25 percent of the elements in terms of abundance in the earth.

The only important ore of cadmium is greenockite, or cadmium sulfide (CdS). Most cadmium is obtained as a by-product of zinc refinement.

As of 2008, China, Korea, Canada, Kazakhstan, Japan, Mexico, Russia, and the United States were among the largest producers of cadmium.

Isotopes

Seven naturally occurring isotopes of cadmium exist. They are cadmium-106, cadmium-108, cadmium-110, cadmium-111, cadmium-112, cadmium-114, and cadmium-116. Isotopes are two or more forms of an element. Isotopes differ from each other according to their mass number. The number written to the right of the element's name is the mass number. The mass number represents the number of protons plus neutrons in the nucleus of an atom of the element. The number of protons determines the element, but the number of neutrons in the atom of any one element can vary. Each variation is an isotope.

Thirty-five radioactive isotopes of cadmium are known also. A radioactive isotope is one that breaks apart and gives off some form of radiation. Radioactive isotopes are produced when very small particles are fired at atoms. These particles stick in the atoms and make them radioactive.

One isotope of cadmium, cadmium-109, is sometimes used to analyze metal alloys. It provides a way of keeping track of the alloys in stock and sorting different forms of scrap metal from each other.

Extraction

Most cadmium is obtained as a by-product from zinc refinement. Cadmium and zinc melt at different temperatures, providing one way of separating the two metals. As a liquid mixture of zinc and cadmium is cooled, zinc becomes a solid first. It can be removed from the mixture, leaving liquid cadmium behind.

In 2008, American consumption of cadmium metal totaled approximately $3.87 million.

Uses

At one time, the most important use of cadmium was in the electroplating of steel. Electroplating is a process by which a thin layer of one metal is deposited on the surface of a second metal. An electric current is passed through a solution containing the coating metal. The metal is electrically deposited on the second metal. A thin layer of cadmium protects steel from corrosion (rusting).

Since the 1960s, the use of cadmium for electroplating has dropped significantly due to environmental concerns. Discarded electroplated steel puts cadmium into the environment. Alternative coating methods are usually used now.

Much of the cadmium produced worldwide is used in nickel-cadmium (nicad or Ni-Cd) batteries. Nicad batteries can be used over and over. When a nicad battery has lost some or all of its power, it is inserted into a unit that plugs into an electrical outlet. Electricity from the outlet recharges the battery.

Nicad batteries are used in a large variety of appliances, including compact disc players, cell phones, pocket recorders, handheld power tools, cordless telephones, laptop computers, camcorders, and scanner radios. A few automobile manufacturers have explored the possibility of using nicad batteries in electric cars. For example, the Mitsubishi Corporation of Japan built and marketed an experimental car called the Libro EV that operated on either traditional lead batteries or on nicad batteries.

Nicad (nickel-cadmium) batteries can be recharged in a plug-in unit such as this one. IMAGE COPYRIGHT 2009, ~WETC~. USED UNDER LICENSE FROM SHUTTERSTOCK.COM.

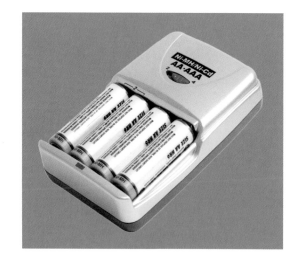

Compounds

A popular use of cadmium compounds is as coloring agents. The two compounds most commonly used are cadmium sulfide (CdS) and cadmium selenide (CdSe). The sulfide is yellow, orange, or brown, while the selenide is red. These compounds are used to color paints and plastics. There is concern about possible environmental effects of using cadmium for this purpose, and scientists have been attempting to find substitutes for these compounds. Some satisfactory alternatives have been found.

Health Effects

Cadmium enters the human body as the result of cigarette smoking, eating certain foods (such as shellfish, liver, and kidney meats), coal burning, and drinking contaminated water. Those most at risk for high intake of cadmium are people who work directly with the metal. Manufacturing plants where batteries are made use cadmium as a fine powder where it can easily be inhaled. Workers must be careful in handling cadmium.

There is growing concern about the dangers of cadmium in the environment. Some rechargeable batteries are made with cadmium and nickel. Cadmium can escape from landfills (where trash is buried) and get into the ground and groundwater. From there, it can become part of the food and water that humans and other animals ingest.

Low levels of cadmium cause nausea, vomiting, and diarrhea. Inhaled, cadmium dust causes dryness of the throat, choking, headache, and pneumonia-like symptoms.

The effects of extensive cadmium exposure are not known, but are thought to include heart and kidney disease, high blood pressure, and cancer. A cadmium poisoning disease called *itai-itai,* Japanese for "ouch-ouch," causes aches and pains in the bones and joints. In the 1970s, a number of cases of *itai-itai* were reported in Japan when waste from a zinc refinery got into the public water supply. Those wastes contained cadmium compounds.

In 2010, product testing revealed that high levels of cadmium were being used in some children's jewelry made in Asia. In response, various U.S. stores pulled those products off their shelves, and a recall was initiated.

Calcium

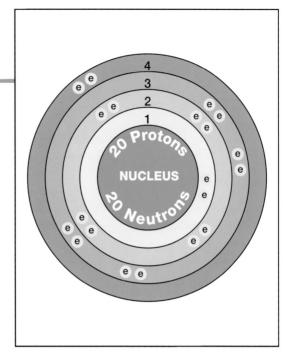

Overview

Calcium is an alkaline earth metal. The alkaline earth metals make up Group 2 (IIA) of the periodic table, a chart that shows how the elements are related. They also include **beryllium**, **magnesium**, **strontium**, **barium**, and **radium**. The alkaline earth metals are more chemically active than most metals. Only the alkali metals in Group I (IA) are more reactive.

Calcium compounds are common and abundant in Earth's crust. Humans have used calcium compounds for thousands of years in construction, sculpture, and roads.

Calcium metal was not prepared in a pure form until 1808 when English chemist Humphry Davy (1778–1829) passed an electric current through molten (melted) calcium chloride.

Metallic calcium has relatively few uses. However, calcium compounds are well known and widely used. They include chalk, gypsum, limestone, marble, and plaster of paris.

Discovery and Naming

It is impossible to say when humans first knew about or used compounds of calcium. Whenever they used limestone to build a structure, they were

Key Facts

Symbol: Ca

Atomic Number: 20

Atomic Mass: 40.078

Family: Group 2 (IIA); alkaline earth metal

Pronunciation: CAL-cee-um

WORDS TO KNOW

Alkaline earth metal: An element found in Group 2 (IIA) of the periodic table.

Alloy: A mixture of two or more metals with properties different from those of the individual metals.

Calx: The original term for calcium.

Isotopes: Two or more forms of an element that differ from each other according to their mass number.

Periodic table: A chart that shows how the chemical elements are related to each other.

Quarry: A large hole in the ground from which useful minerals are taken.

Radioactive isotope: An isotope that breaks apart and gives off some form of radiation.

Tracer: A radioactive isotope whose presence in a system can easily be detected.

Humphry Davy. LIBRARY OF CONGRESS.

using a compound of calcium. Limestone is the common name for calcium carbonate ($CaCO_3$). Whenever humans built a statue or monument out of marble, they were using calcium carbonate in another form. Ancient Egyptians and early Greeks used mortar, a cement-like material that holds stones and bricks together. Early mortar was made by roasting or heating limestone for long periods of time. Water was then mixed with the powder, which would then dry to form a strong bond.

Another calcium compound used by early civilizations was plaster of paris. Plaster of paris is made by heating gypsum, or calcium sulfate ($CaSO_4$), to remove the water that makes it crystallize. Water was added and it hardened into a brittle, cement-like substance. Until recently, it was most often used to make casts to protect broken bones. However, it has largely been replaced by fiberglass, which is lighter, yet stronger. The first mention of plaster of paris to protect broken bones can be found in a book written by Persian pharmacist Abu Mansur Muwaffaw in about 975 CE.

By the 1700s, chemists had learned a great deal about calcium compounds. They knew that limestone, gypsum, marble, and many other commonly occurring compounds all contain a

Element Discoverer: Humphry Davy

Born in 1778, Humphry Davy grew up in Cornwall, England, in a poor family. His father, who died when Davy was a boy, had lost money in unwise investments. So Davy worked to help his mother pay off the debts. He disliked being a student, although he liked reading about science.

With no money for further education, Davy began to work for a surgeon-pharmacist. He also started learning about geography, languages, and philosophy on his own. Davy even dabbled in poetry. At 19, he decided to concentrate on chemistry and eventually became a major contributor to the field of electrochemistry, the science involving the relation of electricity to chemical changes.

Davy discovered nitrous oxide after testing the effects of hydrogen and carbon dioxide on himself. (He liked to use himself as a human guinea pig.) Nitrous oxide is a gas that consists of nitrogen and oxygen. He discovered that its effects often made him feel very happy or very sad. The feeling of happiness eventually gave nitrous oxide another name: laughing gas. Most importantly, Davy recognized that it could be used as an anesthetic. An anesthetic is a chemical used to dull pain during surgery.

In 1808, Davy invented the carbon arc lamp. He had proposed using carbon as the electrode material instead of metal. (Electrodes are conductors used to establish electrical contact with a nonmetallic part of a circuit.) With carbon electrodes, he made a strong electric current leap from one electrode to the other. This created an intense white light. Davy's invention marked the beginning of the era of electric light. Arc lamps are still used today.

In addition, Davy built a large battery that he used to break down substances that most scientists thought were pure elements. In 1807, he discovered the element potassium by using a process known as electrolysis. Electrolysis is a reaction in which electric current is used to bring about chemical changes. Less than a week later, Davy also isolated the element sodium by the same procedure. Then in 1808, he isolated calcium, magnesium, barium, and strontium. Davy was only 29 by the time he had discovered all of these elements.

Davy went on to make other major discoveries and inventions. During his lifetime, Davy was awarded many honors and medals. He died of a stroke in 1829.

common element. They called the element calx. That word comes from the Latin term for lime. In 1807, Davy isolated the new element.

Davy invented a method for extracting elements from compounds that were difficult to separate by usual methods. He first passed an electric current through the compound, causing it to melt. The electric current then caused the compound to break apart into the elements of which it is made.

One of the compounds he used this method on was calx (calcium oxide; CaO), producing pure calcium metal for the first time. Davy suggested the name calcium for the new element. He chose the name by adding the suffix *-ium* to the word calx; *-ium* is the ending used for almost all metallic elements. Using the same method, Davy was also able to produce free **sodium**, **potassium**, strontium, magnesium, and barium.

Physical Properties

Calcium is a fairly soft metal with a shiny silver surface when first cut. The surface quickly becomes dull as calcium reacts with oxygen to form a coating of white or gray calcium oxide.

Calcium's melting point is 1,560°F (850°C) and its boiling point is 2,620°F (1,440°C). It has a density of 1.54 grams per cubic centimeter.

Chemical Properties

Calcium is a moderately active element. It reacts readily with oxygen to form calcium oxide (CaO):

$$2Ca + O_2 \rightarrow 2CaO$$

Calcium reacts with the halogens—**fluorine**, **chlorine**, **bromine**, **iodine**, and **astatine**. The halogens are the elements that make up Group 17 (VIIA) of the periodic table. Calcium also reacts readily with cold water, most acids, and many non-metals, such as **sulfur** and **phosphorus**. For example, calcium reacts with sulfur:

$$Ca + S \rightarrow CaS$$

Occurrence in Nature

Calcium is the fifth most common element in Earth's crust. Its abundance is estimated to be about 3.64 percent. It is also the fifth most abundant element in the human body.

Calcium does not occur as a free element in nature. It is much too active and always exists as a compound. The most common calcium compound is calcium carbonate ($CaCO_3$). It occurs in the minerals aragonite, calcite, chalk, limestone, marble, and travertine, and in oyster shells and coral.

Shellfish build their shells from calcium dissolved in the water. When the animals die or are eaten, the shells sink. Over many centuries, thick layers of the shells may build up and be covered with mud, sand, or other materials. The shells are squeezed together by the heavy pressure of other materials and water above them. As they are squeezed together, the layer is converted to limestone. If the limestone is squeezed even more, it can change into marble or travertine.

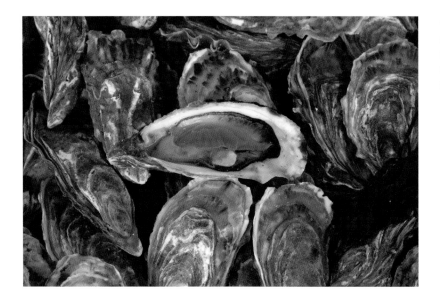

Isotopes

Five naturally occurring stable isotopes of calcium exist: calcium-40, calcium-42, calcium-43, calcium-44, and calcium-46. Isotopes are two or more forms of an element. Isotopes differ from each other according to their mass number. The number written to the right of the element's name is the mass number. The mass number represents the number of protons plus neutrons in the nucleus of an atom of the element. The number of protons determines the element, but the number of neutrons in the atom of any one element can vary. Each variation is an isotope.

Fourteen radioactive isotopes of calcium are also known, one of which, calcium-48, exists in nature. A radioactive isotope is one that breaks apart and gives off some form of radiation. Radioactive isotopes are produced when very small particles are fired at atoms. These particles stick in the atoms and make them radioactive.

Two radioactive isotopes of calcium are used in research and medicine. Calcium-45 is used to study how calcium behaves in many natural processes. For example, it can be used to see how various types of soil behave with different kinds of fertilizers. The calcium-45 is used as a tracer in such studies. A tracer is a radioactive isotope whose presence in a system can easily be detected. The isotope is injected into the system at some point. Inside the system, the isotope gives off radiation. The radiation can be followed by detectors placed around the system.

Calcium-45 can also be used as a tracer in the study of glassy materials, detergents, and water purification systems.

Both calcium-45 and calcium-47 can be used to study how calcium is used in the body. A doctor may think that a person's body is not using calcium properly in making bones or regulating nerve messages. The doctor can use calcium-45 or calcium-47 to find out more about this problem. The radioactive isotope is injected into the person's bloodstream. Then its path can be followed by the radiation it gives off. The doctor can then tell if the calcium is being used normally in the body.

Extraction

Pure calcium metal can be made by the same method used by Davy. An electric current is passed through molten calcium chloride:

$$CaCl_2 \xrightarrow{\text{electric current}} Ca + Cl_2$$

There is not much demand for pure calcium. Most calcium is used in the form of limestone, gypsum, or other minerals that can be mined directly from the earth.

Uses

Calcium metal has relatively few uses. It is sometimes used as a "getter." A getter is a substance that removes unwanted chemicals from a system. Calcium is used as a getter in the manufacture of evacuated glass bulbs. Calcium is added to the bulb while it is being made. It then combines with gases left in the glass in the final stages of manufacture. Calcium is also used as a getter in the production of certain metals, such as **copper** and steel. The calcium removes unwanted elements that would otherwise contaminate the metal.

Calcium is also used to make alloys. An alloy is made by melting and mixing two or more metals. The mixture has properties different from those of the individual metals. An alloy of calcium and **cerium** is used in flints found in lighters (the parts of a lighter that create sparks).

Compounds

The starting point for the manufacture of most calcium compounds is limestone. Limestone occurs naturally in large amounts in many parts

of the world. It is usually mined from open-pit quarries. A quarry is a large hole in the ground from which useful minerals are taken.

Limestone is first heated to obtain lime, or calcium oxide (CaO):

$$\textbf{CaCO}_3 \text{ —heated} \rightarrow \textbf{CaO (lime)} + \textbf{CO}_2$$

Lime is one of the most important chemicals in the world. It usually ranks in the top 10 chemicals produced in the United States.

Lime is used in the production of metals. It is used during the manufacture of steel to remove unwanted sand, or silicon dioxide (SiO$_2$), present in iron ore:

$$\textbf{CaO} + \textbf{SiO}_2 \rightarrow \textbf{CaSiO}_3$$

The product formed in this reaction, calcium silicate (CaSiO$_3$), is called slag.

Another important use of lime is in pollution control. Many factories release harmful gases into the atmosphere through smokestacks. Lining a smokestack with lime allows some of these gases to be captured. The lime is known as a scrubber. Lime captures one harmful gas, sulfur dioxide (SO$_2$), which is a contributor to acid rain (a form of precipitation that is significantly more acidic than neutral water, often produced as the result of industrial processes):

$$\textbf{CaO} + \textbf{SO}_2 \rightarrow \textbf{CaSO}_3$$

Calcium sulfite ($CaSO_3$) is a solid that can be removed from the inside of the smokestack.

Lime is also used in water purification and waste treatment plants. When calcium oxide combines with water, it forms slaked lime, or calcium hydroxide ($Ca(OH)_2$):

$$CaO + H_2O \rightarrow Ca(OH)_2$$

Slaked lime traps impurities present in the water as it forms. It carries the impurities with it as it sinks to the bottom of the tank.

At one time, lime was used as a source of light in theaters. When lime is heated to a high temperature, it gives off an intense white light. Pots of hot lime were often used to line the front of the stage. The light the pots gave off helped the audience see the performers. As a result, the performers were said to be "in the limelight." That phrase is still in use today, but lime is no longer used as a source of light in theaters.

Lime is used to make more than 150 different industrial chemicals. Some examples of these chemicals with their uses are:

- calcium alginate: thickening agent in food products such as ice cream and cheese products; synthetic fibers
- calcium arsenate ($Ca_3(AsO_4)_2$): insecticide
- calcium carbide (CaC_2): used to make acetylene gas (for use in acetylene torches for welding); manufacture of plastics
- calcium chloride ($CaCl_2$): ice removal and control of dust on dirt roads; conditioner for concrete; additive for canned tomatoes; provides body for automobile and truck tires
- calcium cyclamate ($Ca(C_6H_{11}NHSO_4)_2$): sweetening agent (cyclamate), no longer permitted for use in the United States because of suspected cancer-causing properties
- calcium gluconate ($Ca(C_6H_{11}O_7)_2$): food additive; vitamin pills
- calcium hypochlorite ($Ca(OCl)_2$): swimming pool disinfectant; bleaching agent; deodorant; algicide and fungicide (kills algae and fungi)
- calcium permanganate ($Ca(MnO_4)_2$): liquid rocket propellant; textile production; water sterilizing agent; dental procedures
- calcium phosphate ($Ca_3(PO_4)_2$): supplement for animal feed; fertilizer; commercial production of dough and yeast products; manufacture of glass; dental products

- calcium phosphide (Ca_3P_2): fireworks; rodenticide (kills rats); torpedoes; flares
- calcium stearate ($Ca(C_{18}H_{35}O_2)_2$): manufacture of wax crayons, cements, certain kinds of plastics, and cosmetics; food additive; production of water resistant materials; production of paints
- calcium tungstate ($CaWO_4$): luminous paints; fluorescent lights; X-ray studies in medicine.

Health Effects

Calcium is essential to both plant and animal life. In humans, it makes up about 2 percent of body weight. About 99 percent of the calcium in a person's body is found in bones and teeth. Milk is a good source of calcium. The body uses calcium in a compound known as hydroxyapatite ($Ca_{10}(PO_4)_6(OH)_2$) to make bones and teeth hard and resistant to wear.

Calcium has many other important functions in the human body. For example, it helps control the way the heart beats. An excess (too much) or deficiency (not enough) of calcium can change the rhythm of the heart and cause serious problems. Calcium also controls the function of other muscles and nerves.

Californium

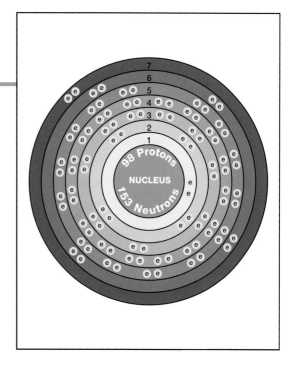

Overview

Californium is a transuranium element, or "beyond **uranium**" on the periodic table. The periodic table is a chart that shows how chemical elements are related to each other. Uranium is element number 92 in the periodic table, so elements with atomic numbers greater than 92 are said to be transuranium elements.

Discovery and Naming

Californium was discovered in 1950 by a research team at the University of California at Berkeley. The team—made up of Glenn Seaborg (1912–1999), Albert Ghiorso (1915–), Kenneth Street Jr. (1920–2006), and Stanley G. Thompson (1912–1976)—named the new element after the state of California.

Californium was first prepared in a particle accelerator, or an "atom smasher," which accelerates subatomic particles or atoms to very high speeds. The particles collide with a target, such as **gold**, **copper**, or **tin**. The target atoms are converted into new elements by the interaction.

To make californium, researchers fired alpha particles (**helium** atoms without electrons) at a target of **curium**. Some collisions cause a helium

Key Facts

Symbol: Cf

Atomic Number: 98

Atomic Mass: [251]

Family: Actinoid; transuranium element

Pronunciation: cal-uh-FOR-nee-um

WORDS TO KNOW

Actinoid family: Elements with atomic numbers 89 through 103.

Half life: The time it takes for half of a sample of a radioactive element to break down.

Isotopes: Two or more forms of an element that differ from each other according to their mass number.

Periodic table: A chart that shows how the chemical elements are related to each other.

Radioactive: Having a tendency to give off radiation.

Transuranium element: An element with an atomic number greater than 92.

atom (atomic number 2) to become part of a curium atom (atomic number 96), forming a new atom with atomic number 98.

Physical and Chemical Properties

Very little is known about the physical and chemical properties of californium. Its melting point has been found to be 1,652°F (900°C) and its density, 15.1 grams per cubic centimeter, about 15 times that of water. It is also very radioactive. One microgram (millionth of a gram) of the element emits about three million neutrons per second.

Albert Ghiorso was part of the team that discovered Californium. AP IMAGES.

Occurrence in Nature

Californium does not occur naturally on Earth. However, it has been observed in the spectra of supernovae.

Isotopes

All 21 isotopes of californium are radioactive. The most stable isotope is californium-251. Isotopes are two or more forms of an element. Isotopes differ from each other according to their mass number. The number written to the right of the element's name is the mass number. The mass number represents the number of protons plus neutrons in the nucleus of an atom of the element. The number of protons determines the element, but the number of neutrons in the atom of any one element can vary. Each variation is an isotope. A radioactive isotope is one that breaks apart and gives off some form of radiation.

The half life of californium-251 is 898 years. The half life of a radioactive element is the time it takes for half of a sample of the element to break down. For instance, in 898 years, only half of a 100-gram sample of californium-251 would remain. After another 898 years, only half of that amount (25 grams) would remain.

One isotope of californium is of special interest: californium-252. This isotope has the unusual property of giving off neutrons when it breaks apart. Isotopes that behave in this way are somewhat rare.

Extraction

Californium has not been observed naturally on Earth.

Uses

When neutrons collide with an atom, they tend to become part of the nucleus, making the atom less stable:

neutron from californium + ordinary copper → radioactive copper

The radioactive copper then gives off radiation or energy and particles that can be measured.

Based on this property, californium-252 has been used to prospect for oil and to test materials without breaking them apart or destroying them. The isotope can also be used to determine the amount of moisture in soil, information that is very important to road builders and construction companies. Neutrons from californium-252 can be used to inspect

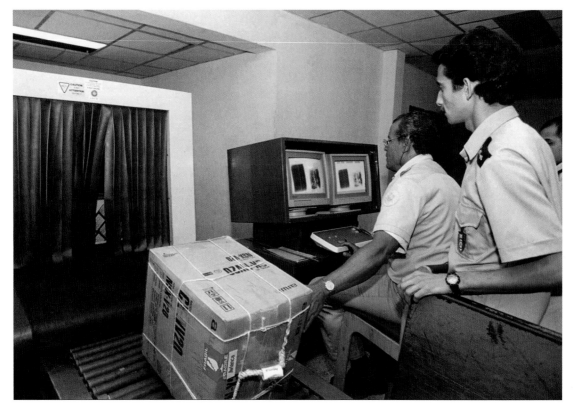

Security personnel use equipment that features californium-252 to inspect luggage. AP IMAGES.

airline baggage. The luggage can be tested quickly and efficiently without having to open it.

Californium-252 is also used in medicine. When injected into the body, it tends to be deposited in bones. The radiation it gives off can be used to determine the health of the bone. Californium-252 is also used to treat ovarian and cervical cancer. Some experts see a number of important medical uses for californium-252 in the future, especially in radiotherapy (medical treatment using radiation).

Today, californium can be made only in small amounts. It is available from the U.S. government via the Oak Ridge National Laboratory in Tennessee.

Compounds

There are no commercially important compounds of californium.

Health Effects

Radioactive materials, such as californium, are hazardous to living cells. As the element's atoms decay, they emit energy and particles that damage or kill the cell. The damaged cells rapidly divide, producing masses called tumors. Cancerous cells can crowd out healthy cells, reduce or stop organ function, and break free to spread through the body.

Carbon

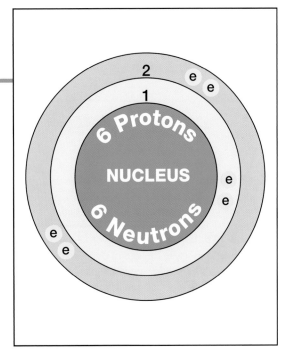

Overview

Carbon is an extraordinary element. It occurs in more compounds than any other element in the periodic table. The periodic table is a chart that shows how chemical elements are related to each other. More than 10 million compounds of carbon are known. No other element, except for **hydrogen**, occurs in even a fraction of that number of compounds.

As an element, carbon occurs in a striking variety of forms. Coal, soot, and diamonds are all nearly pure forms of carbon. Carbon also occurs in a form known as fullerenes or buckyballs. Buckyball carbon holds the promise for opening a whole new field of chemistry.

Carbon occurs extensively in all living organisms as proteins, fats, carbohydrates (sugars and starches), and nucleic acids.

Carbon is such an important element that an entirely separate field of chemistry is devoted to this element and its compounds. Organic chemistry is the study of carbon compounds.

Discovery and Naming

Humans have been aware of carbon since the earliest of times. When cave people made a fire, they saw smoke form. The black color of smoke is

Key Facts

Symbol: C

Atomic Number: 6

Atomic Mass: 12.0107

Family: Group 14 (IVA); carbon

Pronunciation: CAR-bun

WORDS TO KNOW

Amorphous: Lacking crystalline structure.

Biochemistry: The field of chemistry concerned with the study of compounds found in living organisms.

Buckminsterfullerene (buckyball or fullerene): An allotrope of carbon whose 60 carbon atoms are arranged in a sphere-like form.

Carbon-14 dating: A technique that allows archaeologists to estimate the age of once-living materials by using the knowledge that carbon-14 is found in all living carbon materials and that once an organism dies, no more carbon-14 remains.

Global warming: A phenomenon in which the average temperature of Earth rises, melting icecaps, raising sea levels, and causing other environmental problems. Causes include human-activities, including heavy emissions of carbon dioxide (CO_2).

Hydrocarbons: Compounds made of carbon and hydrogen.

Isotopes: Two or more forms of an element that differ from each other according to their mass number.

Mohs scale: A way of expressing the hardness of a material.

Nanotubes: Long, thin, and extremely tiny tubes.

Organic chemistry: The study of the carbon compounds.

Periodic table: A chart that shows how the chemical elements are related to each other.

Photosynthesis: The process by which plants convert carbon dioxide and water to carbohydrates (starches and sugars).

Radioactive isotope: An isotope that breaks apart and gives off some form of radiation.

Refractory: A material that can withstand very high temperatures and reflect heat away from itself.

Sublimation: The process by which a solid changes directly to a gas when heated, without first changing to a liquid.

Toxic: Poisonous.

---

caused by unburned specks of carbon. The smoke may have collected on the ceiling of their caves as soot.

Later, when lamps were invented, people used oil as a fuel. When oil burns, carbon is released in the reaction, forming a sooty covering on the inside of the lamp. That form of carbon became known as lampblack. Lampblack was also often mixed with olive oil or balsam gum to make ink. Ancient Egyptians sometimes used lampblack as eyeliner.

One of the most common forms of carbon is charcoal. Charcoal is made by heating wood in the absence of air so it does not catch fire. Instead, it gives off water vapor, leaving pure carbon. This method for producing charcoal was known as early as the Roman civilization (which thrived from about 509 BCE to about 455 CE).

French physicist René Antoine Ferchault Réaumur (1683–1757) believed carbon might be an element. He studied the differences between wrought **iron**, cast iron, and steel. The main difference among these materials, he said, was the presence of a "black combustible material" that he knew was present in charcoal.

Carbon was officially classified as an element near the end of the 18th century. In 1787, four French chemists–Guyton de Morveau, Antoine Laurent Lavoisier, Claude Louis Berhollet, and Antoine François Fourcroy–wrote a book outlining a method for naming chemical substances, *A Method for Chemical Nomenclature.* The name they gave to carbon was *carbone,* which was based on the earlier Latin term for charcoal, *charbon.*

Physical Properties

Carbon exists in a number of allotropic forms. Allotropes are forms of an element with different physical and chemical properties. Two allotropes of carbon have crystalline structures: diamond and graphite. In a crystalline material, atoms are arranged in a neat orderly pattern. Graphite is found in pencil "lead" and ball-bearing lubricants. Among the non-crystalline allotropes of carbon are coal, lampblack, charcoal, carbon black, and coke. Carbon black is similar to soot. Coke is nearly pure carbon formed when coal is heated in the absence of air. Carbon allotropes that lack crystalline structure are amorphous, or without crystalline shape.

The allotropes of carbon have very different chemical and physical properties. For example, diamond is the hardest natural substance known. It has a rating of 10 on the Mohs scale. The Mohs scale is a way of expressing the hardness of a material. It runs from 0 (for talc) to 10 (for diamond). The melting point of diamond is about 6,700°F (3,700°C) and its boiling point is about 7,600°F (4,200°C). Its density is 3.50 grams per cubic centimeter.

On the other hand, graphite is a very soft material. It is often used as the "lead" in lead pencils. It has a hardness of 2.0 to 2.5 on the Mohs scale. Graphite does not melt when heated, but sublimes at about 6,600°F (3,650°C). Sublimation is the process by which a solid changes directly to a gas when heated, without first changing to a liquid. Its density is about 1.5 to 1.8 grams per cubic centimeter. The numerical value for these properties varies depending on where the graphite originates.

The amorphous forms of carbon, like other non-crystalline materials, do not have clear-cut melting and boiling points. Their densities vary depending on where they originate.

Chemical Properties

Carbon does not dissolve in or react with water, acids, or most other materials. It does, however, react with **oxygen**. It burns in air to produce carbon dioxide (CO_2) and carbon monoxide (CO). The combustion (burning) of coal gave rise to the Industrial Revolution (which began about 1750).

Another highly important and very unusual property of carbon is its ability to form long chains. It is not unusual for two atoms of an element to combine with each other. Oxygen (O_2), **nitrogen** (N_2), hydrogen (H_2), **chlorine** (Cl_2), and **bromine** (Br_2) are a few of the elements that can do this. Some elements can make even longer strings of atoms. Rings of six and eight **sulfur** atoms (S_6 and S_8), for example, are not unusual.

Carbon has the ability to make virtually endless strings of atoms. If one could look at a molecule of almost any plastic, for example, a long chain of carbon atoms attached to each other (and to other atoms as well) would be evident. Carbon chains can be even more complicated. Some chains have side chains attached to them.

In other cases, carbon atoms can join together in rings, boxes, or other shapes. There is almost no limit to the size and shape of molecules that can be made with carbon atoms.

Buckyballs are another form of pure carbon. These spheres are made up of exactly 60 linked carbon atoms.

Occurrence in Nature

Carbon is the fifth most common element in the universe, by weight, and the fourth most common element in the solar system. It is the second most common element in the human body after oxygen. About 18 percent of a person's body weight is due to carbon.

Carbon is the 17th most common element in Earth's crust. Its abundance has been estimated to be between 180 and 270 parts per million. It rarely occurs as a diamond or graphite. Both allotropes are formed in the earth over millions of years, when dead plant materials are squeezed together at very high temperatures. Diamonds are usually found hundreds or thousands of feet beneath the earth's surface. Africa has many diamond mines.

Carbon also occurs in a number of minerals. Among the most common of these minerals are the carbonates of **calcium** ($CaCO_3$) and **magnesium** ($MgCO_3$). Carbon also occurs in the form of carbon dioxide (CO_2) in the atmosphere. Carbon dioxide makes up only a small part of the atmosphere (about 300 parts per million), but it is a crucial gas. Plants use carbon dioxide in the atmosphere in the process of photosynthesis. Photosynthesis is the process by which plants convert carbon dioxide and water to carbohydrates (starches and sugars). This process is the source of life on Earth.

Carbon also occurs in coal, oil, and natural gas. These materials are often known as fossil fuels. They get that name because of the way they were formed. They are the remains of plants and animals that lived millions of years ago. When they died, they fell into water or were trapped in mud. Over millions of years, they slowly decayed. The products of that decay process were coal, oil, and natural gas.

Some forms of coal are nearly pure carbon. Oil and natural gas are made primarily of hydrocarbons, which are compounds made of carbon and hydrogen.

Isotopes

Three isotopes of carbon occur in nature: carbon-12, carbon-13, and carbon-14. One of these isotopes, carbon-14, is radioactive. Isotopes are two or more forms of an element. Isotopes differ from each other according to their mass number. The number written to the right of the element's name is the mass number. The mass number represents the number of protons plus neutrons in the nucleus of an atom of the element. The number of protons determines the element, but the number of neutrons in the atom of any one element can vary. Each variation is an isotope.

Ten artificial radioactive isotopes of carbon have also been synthesized. A radioactive isotope is one that breaks apart and gives off some form of radiation. Artificial radioactive isotopes can be made by firing very small particles (such as protons) at atoms. These particles stick in the atoms and make them radioactive.

Carbon-14 has some limited applications in industry. For example, it can be used to measure the thickness of objects, such as sheets of steel.

In this process, a small sample of carbon-14 is placed above the conveyor belt carrying the steel sheet. A detection device is placed below the sheet. The detection device counts the amount of radiation passing through the sheet. If the sheet gets thicker, less radiation gets through. If the sheet gets thinner, more radiation gets through. The detector records how much radiation passes through the sheet. If the amount becomes too high or too low, the conveyor belt is turned off.

How Carbon-14 Dating Works

When an organism is alive, it takes in carbon dioxide from the air around it. Most of that carbon dioxide is made of carbon-12, but a tiny portion consists of carbon-14. So the living organism always contains a very small amount of radioactive carbon, carbon-14. A detector next to the living organism would record radiation given off by the carbon-14 in the organism.

When the organism dies, it no longer takes in carbon dioxide. No new carbon-14 is added, and the old carbon-14 slowly decays into nitrogen. The amount of carbon-14 slowly decreases as time goes on. Over time, less and less radiation from carbon-14 is produced. The amount of carbon-14 radiation detected for an organism is a measure, therefore, of how long the organism has been dead. This method of determining the age of an organism is called carbon-14 dating.

The decay of carbon-14 allows archaeologists (people who study old civilizations) to find the age of once-living materials. Measuring the amount of radiation remaining indicates the approximate age.

The machine making the sheet is adjusted to produce steel of the correct thickness.

The most important use of carbon-14 is in finding the age of old objects (see accompanying sidebar for more information).

Extraction

Diamond, graphite, and other forms of carbon are taken directly from mines in the earth. Diamond and graphite can also be made in laboratories. Synthetic diamonds, for example, are made by placing pure carbon under very high pressures (about 800,000 pounds per square inch/56,000 kilograms per square centimeter) and temperatures (about 4,900°F/2,700°C;). The carbon is heated and squeezed in the same way organic material is heated and squeezed in the earth. Today, about a third of all diamonds used are synthetically produced.

Uses

There are many uses for carbon's two key allotropes, diamond and graphite. Diamonds are one of the most beautiful and expensive gemstones in the world. But they also have many industrial uses. Because they are so hard, they are used to polish, grind, and cut glass, metals, and other materials. The bit on an oil-drilling machine may be made of diamonds. The tool used to make thin **tungsten** wires is also made of diamonds.

Synthetic diamonds are more commonly used in industry than in jewelry. Industrial diamonds do not have to be free of flaws, as do jewelry diamonds.

Graphite works well as pencil lead because it rubs off easily. It is also used as a lubricant. Graphite is added to the space between machine parts that rub against each other. The graphite allows the parts to slide over each other smoothly.

Graphite is also used as a refractory. Refractory material can withstand very high temperatures by reflecting heat away from itself. Refractory materials are used to line ovens to maintain high temperatures.

Graphite is used in nuclear power plants. A nuclear power plant converts nuclear energy to electrical power. Graphite acts as a moderator by slowing down the neutrons used in the nuclear reaction.

Graphite is used to make black paint, in explosives and matches, and in certain kinds of cathode ray tubes, like the ones used in television sets.

Amorphous forms of carbon have many uses. These include the black color in inks, pigments (paints), rubber tires, and dry cells.

One form of carbon is known as activated charcoal. The term "activated" means that the charcoal has been ground into a very fine powder. In this form, charcoal can remove impurities from liquids that pass through. For example, activated charcoal removes color and odor from oils and water solutions.

Buckyballs and Nanotubes In the 1980s, chemists discovered a new allotrope of carbon. The carbon atoms in this allotrope are arranged in a sphere-like form of 60 atoms. The form resembles a building invented by American architect Buckminster Fuller (1895–1983). The building is known as a geodesic dome.

The discoverers named this new form of carbon "buckminsterfullerene" in honor of Fuller. That name is too long to use in everyday conversation so it is usually shortened to fullerene or buckyball.

The discovery of the fullerene molecule was very exciting to chemists. They had never seen a molecule like it. They have been studying ways of working with this molecule. One interesting technique has been to cut open just one small part of the molecule. Then they cut open a small part

on a second molecule. Finally, they join the two buckyballs together. They get a double-buckyball.

Repeating this process over and over could result in triple-buckyballs, quadruple-buckyballs, and so on. As this process is repeated, the buckyball becomes a long narrow tube called a nanotube. Nanotubes are long, thin, and extremely tiny tubes somewhat like a drinking straw or a long piece of spaghetti.

Scientists have discovered ways of using nanotubes. One idea is to run a thin chain of metal atoms through the center of a nanotube. The nanotube then acts like a tiny electrical wire.

Compounds

Carbon dioxide (CO_2) is used in fire extinguishers and as a propellant in aerosol products. A propellant is a gas that pushes liquids out of a spray can,

The fizz in various beverages is caused by carbonation. IMAGE COPYRIGHT 2009, VOLODYMYR KRASYUK. USED UNDER LICENSE FROM SHUTTERSTOCK.COM.

such as those used for deodorant or hair spray. It is also used to make carbonated beverages (it produces the fizz in soda pop and beer). Carbon dioxide can also be frozen to a solid called dry ice, widely used as a way of keeping objects cold.

Carbon monoxide (CO) is another compound formed between carbon and oxygen. Carbon monoxide is a very toxic gas produced when something burns in a limited amount of air. Carbon monoxide is always formed when gasoline burns in the engine of an automobile and is a common part of air pollution. Old heating units can produce carbon monoxide. This colorless and odorless gas can cause headaches, illness, coma, or even death.

Carbon monoxide has a few important industrial uses. It is often used to obtain a pure metal from the ore of that metal:

$$3CO + Fe_2O_3 \rightarrow 3CO_2 + 2Fe$$

It would take a very large book to describe all the uses of organic compounds, which are divided into a number of families. An organic family is a group of organic compounds with similar structures and properties. The largest organic family is the hydrocarbons, compounds that contain only carbon and hydrogen. Methane, or natural gas (CH_4), ethane (C_2H_6), propane (C_3H_8), ethylene (C_2H_4), and benzene (C_6H_6) are all hydrocarbons.

Hydrocarbons are used as fuels. Gas stoves burn natural gas, which is mostly methane. Propane gas is a popular camping fuel, used in small stoves and lanterns. Another important use of hydrocarbons is in the production of more complicated organic compounds.

Other organic families contain carbon, hydrogen, and oxygen. Methyl alcohol (wood alcohol) and ethyl alcohol (grain alcohol) are the most common members of the alcohol family. Methyl alcohol is used to make other organic compounds and as a solvent (a substance that dissolves other substances). Ethyl alcohol is used for many of the same purposes. It is also the alcohol found in beer, wine, and hard liquor, such as whiskey and vodka.

All alcohols are poisonous but some alcohols are more poisonous than others. If not drunk in moderation, alcoholic beverages can damage the body and brain. If consumed in large quantities, they can cause death. Methyl alcohol is more toxic than ethyl alcohol. People who have consumed methyl alcohol by mistake have died.

The list of everyday products made from organic compounds is very long. It includes drugs, artificial fibers, dyes, artificial colors and flavors, food additives, cosmetics, plastics of all kinds, detergents, synthetic rubber, adhesives, antifreeze, pesticides and herbicides, synthetic fuels, and refrigerants.

Health Effects

Carbon is essential to life. Nearly every molecule in a living organism contains carbon. The study of carbon compounds that occur in living organisms is called biochemistry (*bio-* = life + *-chemistry*).

Carbon can also have harmful effects on organisms. For example, coal miners sometimes develop a disease known as black lung. The name comes from the appearance of the miner's lungs. Instead of being pink and healthy, the miner's lungs are black. The black color is caused by coal dust inhaled by the miner. The longer a miner works digging coal, the more the coal dust is inhaled. That worker's lungs become more and more black.

Color is not the problem with black lung disease however. The coal dust in the lungs blocks the tiny holes through which oxygen gets into the lungs. As more coal dust accumulates, more holes are plugged up, making it harder for the miner to breathe. Many miners eventually die from black lung disease because they lose the ability to breathe.

Carbon monoxide poisoning is another serious health problem. Carbon monoxide is formed whenever coal, oil, or natural gas burns. For example, the burning of gasoline in cars and trucks produces carbon monoxide. Today, almost every person in the United States inhales some carbon monoxide every day.

Small amounts of carbon monoxide are not very dangerous. But larger amounts cause a variety of health problems. At low levels, carbon monoxide causes headaches, dizziness, nausea, and loss of balance. At higher levels, a person can lose consciousness. At even higher levels, carbon monoxide can cause death.

Most cars emit exhaust containing CO₂. IMAGE COPYRIGHT 2009, TYLER OLSON. USED UNDER LICENSE FROM SHUTTERSTOCK.COM.

Scientists are actively studying the effects of carbon dioxide (CO_2) on people and the environment. CO_2 is released into the air primarily through the burning of fossil fuels (gas and oil), the logging of forests (clear-cutting), and the use of fuel-powered vehicles. Many climate scientists believe that the release of CO_2 into the atmosphere is causing climate change, including global warming. They are studying the situation to see if the melting of glaciers and icebergs, increased storm activity, and warmer-than-normal temperatures are due to increasing amounts of CO_2 being released into the atmosphere by humans.

Cerium

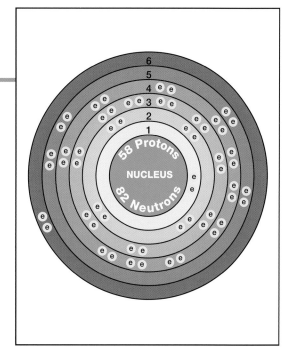

Overview

Cerium is the most abundant of the rare earth metals. Rare earth metals are found in row 6 of the periodic table. The periodic table is a chart that shows how chemical elements are related to each other. The rare earth elements are not really rare. In fact, cerium ranks about number 26 in abundance among elements found in Earth's crust.

Cerium is a gray metal that easily reacts with other elements. It is used in making a number of different alloys, in the production of many kinds of specialty glass, and in the chemical industry.

Discovery and Naming

Cerium was the first rare earth element to be discovered. It was isolated in 1839 by Swedish chemist Carl Gustav Mosander (1797–1858). Mosander was studying a new rock that had been discovered outside the town of Bastnas, Sweden. Mosander named the new element cerium, in honor of the asteroid Ceres that had been discovered in 1801.

Credit for the discovery of cerium is sometimes given to scientists who studied the black rock of Bastnas earlier. These scientists included Swedish chemists Jöns Jakob Berzelius (1779–1848) and Wilhelm

Key Facts

Symbol: Ce

Atomic Number: 58

Atomic Mass: 140.116

Family: Lanthanoid (rare earth metal)

Pronunciation: SEER-ee-um

WORDS TO KNOW

Abrasive: A powdery material used to grind or polish other materials.

Allotropes: Forms of an element with different physical and chemical properties.

Alloy: A mixture of two or more metals with properties different from those of the individual metals.

Catalyst: A substance used to speed up a chemical reaction without undergoing any change itself.

Ductile: Capable of being drawn into thin wires.

Isotopes: Two or more forms of an element that differ from each other according to their mass number.

Lanthanoids: The elements in the periodic table with atomic numbers 57 through 71.

Laser: A device for making very intense light of one very specific color that is intensified many times over.

Malleable: Capable of being hammered into thin sheets.

Periodic table: A chart that shows how chemical elements are related to each another.

Phosphor: A material that gives off light when struck by electrons.

Hisinger (1766–1852) and German chemist Martin Heinrich Klaproth (1743–1817). It would be difficult to say that one or another of these chemists was the one and only discoverer of cerium.

The substance these scientists discovered was not a pure element but cerium combined with oxygen and other elements. Pure cerium was not produced for another 70 years.

Physical Properties

Cerium is an iron-gray metal with a melting point of 1,460°F (795°C) and a boiling point of 5,895°F (3,257°C). It is ductile and malleable. Ductile means capable of being made into thin wires. Malleable means capable of being hammered into thin sheets. Cerium's density is 6.78 grams per cubic centimeter. It exists in four different allotropic forms. Allotropes are forms of an element with different physical and chemical properties.

Chemical Properties

Cerium is the second most active lanthanoid after **europium**. Lanthanoids are the elements with atomic numbers 57 through 71. Cerium reacts so readily with **oxygen** that it can be set on fire simply by scratching the surface with a knife. It also reacts with cold water (slowly), hot water (rapidly), acids, bases, **hydrogen** gas, and other metals. Because it is so active, it must be handled with caution.

Occurrence in Nature

The most important ores of cerium are cerite, monazite, and bastnasite. It is thought to occur in Earth's crust with a concentration of 40 to 66 parts per million. This makes cerium about as abundant as **copper** or **zinc**.

Isotopes

Four naturally occurring isotopes of cerium have been discovered: cerium-136, cerium-138, cerium-140, and cerium-142. The last of these isotopes is radioactive. Isotopes are two or more forms of an element. Isotopes differ from each other according to their mass number. The number written to the right of the element's name is the mass number. The mass number represents the number of protons plus neutrons in the nucleus of an atom of the element. The number of protons determines the element, but the number of neutrons in the atom of any one element can vary. Each variation is an isotope. A radioactive isotope is one that breaks apart and gives off some form of radiation.

Thirty-three radioactive isotopes of cerium have also been made. Radioactive isotopes are produced when very small particles are fired at atoms. These particles stick in the atoms and make them radioactive. None of the radioactive isotopes of cerium has any commercial use.

Extraction

Cerium is prepared by methods similar to those used for other lanthanoids. It is obtained by passing an electric current through cerium chloride:

$$2CeCl_3 \rightarrow 2Ce + 3Cl_2$$

or by heating calcium metal together with cerium fluoride:

$$3Ca + 2CeF_3 \rightarrow 2Ce + 3CaF_2$$

Uses and Compounds

Cerium metal and its compounds have a great variety of uses, many in the field of glass and ceramics. Cerium and its compounds are added to these materials to add color (yellow), remove unwanted color, make glass sensitive to certain forms of radiation, add special optical (light) qualities to glass, and strengthen certain kinds of dental materials.

Important applications are being found for cerium lasers. A laser is a device that produces bright light of a single frequency or color. Cerium lasers contain a crystal made of **lithium**, **strontium**, **aluminum**, and **fluorine**, to which a small amount of cerium is added. A cerium laser produces light in the ultraviolet region. Ultraviolet radiation is not visible, but it is very similar to the blue and violet light our eyes can see. Cerium lasers are used to search for ozone and **sulfur** dioxide, two air pollutants, in the atmosphere.

Cerium compounds are also used in making phosphors. A phosphor is a material that shines when struck by electrons. The color of the phosphor depends on the elements of which it is made. Phosphors that contain cerium compounds produce a red or orange light when struck by electrons.

Cerium is also used in catalytic systems. A catalyst is a substance used to speed up a chemical reaction. The catalyst does not undergo any change itself during the reaction. Compounds of cerium are used in the refining of petroleum. They help break down compounds found in petroleum into simpler forms that work better as fuels.

Misch metal, a cerium alloy, gives off a spark when struck. It is used in the flints of cigarette lighters. IMAGE COPYRIGHT 2009, JAMES A. KOST. USED UNDER LICENSE FROM SHUTTERSTOCK.COM.

Another application of cerium (in the form of cerium oxide) is in internal combustion engines, like the one found in cars. Adding cerium oxide (CeO_2) to the engine's fuel helps the fuel burn more cleanly, producing fewer pollutants.

A number of alloys contain cerium. An alloy is made by melting and mixing two or more metals. The mixture has properties different from those of the individual metals. Perhaps the best known alloy of cerium is misch metal. Misch metal contains a number of different rare earth elements and has the unusual property of giving off a spark when struck. It is used, for example, in the flint of a cigarette lighter.

Cerium oxide is also used as an abrasive. An abrasive is a powdery material used to grind or polish other materials. Cerium oxide has replaced an older abrasive known as rouge for polishing specialized glass, such as telescope mirrors.

Health Effects

There is no evidence that cerium compounds pose a health hazard to humans.

Cesium

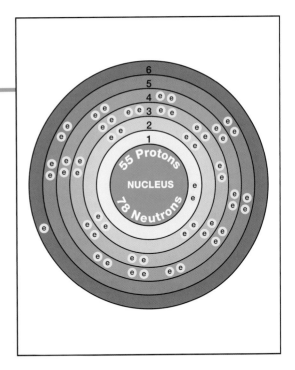

Overview

Cesium is a member of the alkali family, which consists of elements in Group 1 (IA) of the periodic table. The periodic table is a chart that shows how chemical elements are related to each other. The alkalis also include **lithium**, **sodium**, **potassium**, **rubidium**, and **francium**. Cesium is considered the most active metal. Francium is the most active of the alkali elements. It is a very rare element, however, and has few commercial uses.

Cesium was discovered in 1861 by German chemists Robert Bunsen (1811–1899) and Gustav Kirchhoff (1824–1887). They found the element using a method of analysis they had just invented: spectroscopy. Spectroscopy is the process of analyzing light produced when an element is heated. The light produced is different for every element. The spectrum (plural: spectra) of an element consists of a series of colored lines.

Cesium is not a common element, and it has few commercial uses. One of its radioactive isotopes, cesium-137, is widely used in a variety of medical and industrial applications.

Key Facts

Symbol: Cs

Atomic Number: 55

Atomic Mass: 132.9054519

Family: Group 1 (IA); alkali metal

Pronunciation: SEE-zee-um

WORDS TO KNOW

Alkali metal: An element in Group 1 (IA) of the periodic table.

Ductile: Capable of being drawn into thin wires.

Halogen: One of the elements in Group 17 (VIIA) of the periodic table.

Isotopes: Two or more forms of an element that differ from each other according to their mass number.

Nuclear fission: The process in which large atoms break apart.

Periodic table: A chart that shows how chemical elements are related to each other.

Radioactive isotope: An isotope that breaks apart and gives off some form of radiation.

Spectroscopy: The process of analyzing light produced when an element is heated.

Discovery and Naming

The invention of spectroscopy gave chemists a powerful new tool. In many cases, the amount of an element present in a sample is too small to detect by most methods of analysis. But the element can be found by spectroscopy. When a substance is heated, elements give off characteristic spectral lines. Using spectroscopy, a chemist can identify the elements by these distinctive lines.

Such was the case with the discovery of cesium. In 1859, Bunsen and Kirchhoff were studying a sample of mineral water taken from a spring. They saw spectral lines for sodium, potassium, lithium, **calcium**, and **strontium**. These elements were already well known.

After Bunsen and Kirchhoff removed all these elements from their sample, they were surprised to find two beautiful blue lines in the spectrum of the "empty" spring water. The water contained an unknown element. Bunsen suggested calling the element cesium, from the Latin word *caesius* for "sky blue." For many years, the name was also spelled caesium.

Physical Properties

Cesium is a silvery-white, shiny metal that is very soft and ductile. Ductile means capable of being drawn into thin wires. Its melting point is 83.3°F (28.5°C). It melts easily in the heat of one's hand, but should never be handled that way. Cesium's boiling point is 1,300°F (705°C), and its density is 1.90 grams per cubic centimeter.

Chemical Properties

Cesium is a very reactive metal. It combines readily with **oxygen** in the air and reacts violently with water. In the reaction with water, **hydrogen** gas is released. Hydrogen gas ignites immediately as a result of the heat given off by the reaction. Cesium must be stored under kerosene or a mineral oil to protect it from reacting with oxygen and water vapor in the air.

Cesium also reacts vigorously with acids, the halogens, **sulfur**, and **phosphorus**.

Occurrence in Nature

The abundance of cesium in Earth's crust has been estimated at about 1 to 3 parts per million. It ranks in the middle of the chemical elements in terms of their abundance in the earth.

Cesium occurs in small quantities in a number of minerals. It is often found in an ore of lithium called lepidolite. The mineral containing the largest fraction of cesium is pollucite ($Cs_4Al_4Si_9O_{26}$). This ore is mined in large quantities in the Canadian province of Manitoba. Pollucite is also known to exist in Maine and South Dakota, among other places. Cesium is also found in small amounts in a mineral called rhodizite that also contains potassium, **beryllium**, aluminum, and **boron**.

German chemist Robert Bunsen. LIBRARY OF CONGRESS.

Isotopes

Only one naturally occurring isotope of cesium is known: cesium-133. Isotopes are two or more forms of an element. Isotopes differ from each other according to their mass number. The number written to the right of the element's name is the mass number. The mass number represents the number of protons plus neutrons in the nucleus of an atom of the element. The number of protons determines the element, but the number of neutrons in the atom of any one element can vary. Each variation is an isotope.

Fifty-two radioactive isotopes of cesium are known also. A radioactive isotope is one that breaks apart and gives off some form of radiation.

Radioactive isotopes are produced when very small particles are fired at atoms. These particles stick in the atoms and make them radioactive.

Cesium-137 One radioactive isotope of cesium is of special importance, cesium-137. It is produced in nuclear fission reactions. Nuclear fission is the process in which large atoms break apart. Large amounts of energy and smaller atoms are produced during fission. The smaller atoms are called fission products. Cesium-137 is a very common fission product.

Nuclear fission is used in nuclear power plants. The heat produced by nuclear fission can be converted into electricity. While this process is going on, cesium-137 is being produced as a by-product. That cesium-137 can be collected and used for a number of applications.

For example, cesium-137 can be used to monitor the flow of oil in a pipeline. In many cases, more than one oil company may use the same pipeline. How does a receiving station know whose oil is coming through the pipeline? One way to solve that problem is to add a little cesium-137 when a new batch of oil is being sent. The cesium-137 gives off radiation. That radiation can be detected easily by holding a detector at the end of the pipeline. When the detector shows the presence of radiation, a new batch of oil has arrived.

This isotope of cesium can also be used to treat some kinds of cancer. One procedure is to fill a hollow steel needle with cesium-137. The needle can then be implanted into a person's body. The cesium-137 gives off radiation inside the body. That radiation kills cancer cells and may help cure the disease.

Cesium-137 is often used in scientific research. For example, cesium tends to stick to particles of sand and gravel. This fact can be used to measure the speed of erosion in an area. Cesium-137 is injected into the ground at some point. Some time later, a detector is used to see how far the isotope has moved. The distance moved tells a scientist how fast soil is being carried away. In other words, it tells how fast erosion is taking place.

Cesium-137 has also been approved for the irradiation of certain foods. The radiation given off by the isotope kills bacteria and other organisms that cause disease. Foods irradiated by this method last longer before beginning to spoil. Wheat, flour, and potatoes are some of the foods that can be preserved by cesium-137 irradiation.

Extraction

Cesium can be obtained in pure form by two methods. In one, calcium metal is combined with fused (melted) cesium chloride:

$$Ca + 2CsCl \rightarrow 2Cs + CaCl_2$$

In the other, an electric current passes through a molten (melted) cesium compound:

$$2CsCl \xrightarrow{electrical\ current} 2Cs + Cl_2$$

Uses

Cesium has a limited number of uses. One is as a getter in bulbs and evacuated tubes. The bulb must be as free from gases as possible to work properly. Small amounts of cesium react with any air left in the bulb. It converts the gas into a solid cesium compound. Cesium is called a getter because it gets gases out of the bulb.

Cesium is also used in photoelectric cells, devices for changing sunlight into electrical energy. When sunlight shines on cesium, it excites or energizes the electrons in cesium atoms. The excited electrons easily flow away, producing an electric current.

An important use of cesium is in an atomic clock. An atomic clock is the most precise method now available for measuring time. Here is how an atomic clock works:

A beam of energy is shined on a very pure sample of cesium-133. The atoms in the cesium are excited by the energy and give off radiation. That radiation vibrates back and forth, the way a violin string vibrates when plucked. Scientists measure the speed of that vibration. The second is officially defined as that speed of vibration multiplied by 9,192,631,770.

Atomic clocks keep very good time. The best of them lose no more than one second in 60 million years.

Compounds

Cesium compounds have relatively few commercial uses. Cesium bromide is used to make radiation detectors and other measuring devices. Cesium carbonate and cesium fluoride are used to make specialty glasses. Cesium carbonate and cesium chloride are used in the brewing of beers. Cesium compounds are also used in chemical research.

Health Effects

Cesium is not regarded as essential to the health of plants or animals, nor does it present a hazard to them.

Chlorine

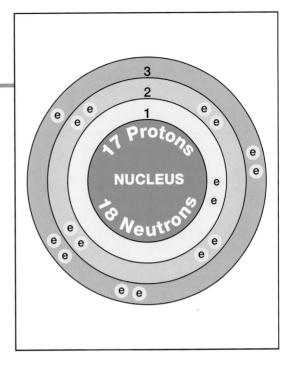

Overview

Chlorine ranks among the top 10 chemicals produced in the United States. In 2008 approximately 11.5 million short tons (10.4 million metric tons) of chlorine were produced in the United States. Chlorine, in one form or another, is added to many swimming pools, spas, and public water supplies because it kills bacteria that cause disease. Many people also use chlorine to bleach their clothes. Large paper and pulp mills use chlorine to bleach their products.

Chlorine is a greenish-yellow poisonous gas. It was discovered in 1774 by Swedish chemist Carl Wilhelm Scheele (1742–1786). Scheele knew that chlorine was a new element, but thought it contained **oxygen** as well.

Chlorine is a member of the halogen family. Halogens are the elements that make up Group 17 (VIIA) of the periodic table, a chart that shows how elements are related to one another. They also include **fluorine**, **bromine**, **iodine**, and **astatine**. Chlorine is highly reactive, ranking only below fluorine in its chemical activity.

Discovery and Naming

Chlorine compounds have been important to humans for thousands of years. Ordinary table salt, for example, is sodium chloride (NaCl). Still, chlorine

Key Facts

Symbol: Cl

Atomic Number: 17

Atomic Mass: 35.453

Family: Group 17 (VIIA); halogen

Pronunciation: CLOR-een

WORDS TO KNOW

Chlorofluorocarbons (CFCs): A family of chemical compounds once used as propellants in commercial sprays but now regulated because of their harmful environmental effects.

Halogen: One of the elements in Group 17 (VIIA) of the periodic table.

Isotopes: Two or more forms of an element that differ from each other according to their mass number.

Micronutrient: A substance needed in very small amounts to maintain good health.

Noble gas: An element in Group 18 (VIIIA) of the periodic table.

Oxidizing agent: A chemical substance that takes on electrons from another substance.

Ozone: A form of oxygen that filters out harmful radiation from the sun.

Periodic table: A chart that shows how chemical elements are related to each other.

Radioactive isotope: An isotope that breaks apart and gives off some form of radiation.

Salt dome: A large mass of salt found underground.

Toxic: Poisonous.

Tracer: An isotope whose presence in a material can be traced (followed) easily.

was not recognized as an element until 1774, when Scheele was studying the mineral pyrolusite. Pyrolusite consists primarily of **manganese** dioxide (MnO_2). Scheele mixed pyrolusite with hydrochloric acid (HCl), then called *spiritus salis.* He found that a greenish-yellow gas with a suffocating odor "most oppressive to the lungs" was released. The gas was chlorine.

Scheele found that the new gas reacted with metals, dissolved slightly in water, and bleached flowers and leaves. He gave the gas the rather complex name of dephlogisticated marine acid.

The true nature of Scheele's discovery was not completely understood for many years. Some chemists argued that his dephlogisticated marine acid was really a compound of a new element and oxygen. This confusion was finally cleared up in 1807. English chemist Sir Humphry Davy (1778–1829) proved that Scheele's substance was a pure element. He suggested the name chlorine for the element, from the Greek word *chloros,* meaning "greenish-yellow."

Physical Properties

Chlorine is a dense gas with a density of 3.21 grams per liter. By comparison, the density of air is 1.29 grams per liter. Chlorine changes from a gas into a liquid at a temperature of −29.29°F (−34.05°C) and from a liquid to a solid at −149.80°F (−101°C). The gas is soluble (dissolvable) in water. It also reacts chemically with water as it dissolves to form hydrochloric acid (HCl) and hypochlorous acid (HOCl).

Chemical Properties

Chlorine is a very active element. It combines with all elements except the noble gases. The noble gases are the elements that make up Group 18 (VIIIA) of the periodic table. The reaction between chlorine and other elements can often be vigorous. For example, chlorine reacts explosively with **hydrogen** to form hydrogen chloride:

$$H_2 + Cl_2 \rightarrow 2HCl$$

Chlorine does not burn but, like oxygen, it helps other substances burn. Chlorine is a strong oxidizing agent (a chemical substance that takes on electrons from another substance).

Swedish pharmacist Carl Wilhelm Scheele. LIBRARY OF CONGRESS.

Occurrence in Nature

Chlorine occurs commonly both in Earth's crust and in seawater. Its abundance in the earth is about 100 to 300 parts per million. It ranks 20th among the elements in abundance in the earth. Its abundance in seawater is about 2 percent. The most common compound of chlorine in seawater is sodium chloride. Smaller amounts of potassium chloride also occur in seawater.

The most common minerals of chlorine are halite, or rock salt (NaCl), sylvite (KCl), and carnallite (KCl • MgCl$_2$). Large amounts of these minerals are mined from underground salt beds that were formed when ancient oceans dried up. Over millions of years, the salts that remained behind were buried underground. They were also compacted (packed together) to form huge salt "domes." A salt dome is a large mass of salt found underground.

Isotopes

Two naturally occurring isotopes of chlorine exist: chlorine-35 and chlorine-36. Isotopes are two or more forms of an element. Isotopes differ

Halite (NaCl) is a common mineral of chloride. IMAGE COPYRIGHT 2009, K–MIKE. USED UNDER LICENSE FROM SHUTTERSTOCK.COM.

from each other according to their mass number. The number written to the right of the element's name is the mass number. The mass number represents the number of protons plus neutrons in the nucleus of an atom of the element. The number of protons determines the element, but the number of neutrons in the atom of any one element can vary. Each variation is an isotope.

Sixteen radioactive isotopes of chlorine are known also. A radioactive isotope is one that breaks apart and gives off some form of radiation. Radioactive isotopes are produced when very small particles are fired at atoms. These particles stick in the atoms and make them radioactive.

One radioactive isotope of chlorine is used in research. That isotope is chlorine-36. This isotope is used because compounds of chlorine occur so commonly in everyday life. The behavior of these compounds can be studied if chlorine-36 is used as a tracer. A tracer is an isotope whose presence in a material can be traced (followed) easily.

For example, engineers are interested in knowing how seawater damages metals. This information is important in determining the best techniques to use in building ships. An experiment can be done by adding pieces of metal to seawater that contains radioactive chlorine-36. The sodium chloride in the seawater is changed slightly so that it contains radioactive chlorine instead of normal chlorine. As the sodium

chloride attacks the metal, its actions can be followed easily. The radioactive chlorine, chlorine-36, gives off radiation. That radiation can be detected by holding an instrument near the experiment. A scientist can find out exactly what happens when the sodium chloride attacks the metal.

Extraction

Chlorine is produced by passing an electric current through a water solution of sodium chloride or through molten (melted) sodium chloride. This process is one of the most important commercial processes in industry. The products formed in the first process include two of the most widely used materials: sodium hydroxide (NaOH) and chlorine (Cl_2). With a water solution, the reaction that occurs is:

$$2NaCl + 2H_2O \xrightarrow{\text{electrical current}} 2NaOH + Cl_2 + H_2$$

Hydrogen gas (H_2) is also formed in the reaction.

Uses and Compounds

Chlorine is widely used throughout the world to purify water. In the United States, about 4 percent of the chlorine manufactured is used in water purification. Very small amounts of chlorine (about 1 percent of all the chlorine produced) is used in the paper and pulp industry as a bleach.

The use of chlorine by the paper and pulp industry has decreased dramatically in the past few decades because of concerns about the effects of the element on the environment.

The most important use of chlorine is to make other chemicals. For example, chlorine can be combined with ethene, or ethylene, gas (C_2H_2), to make ethylene dichloride ($C_2H_2Cl_2$):

$$Cl_2 + C_2H_2 \rightarrow C_2H_2Cl_2$$

Much of the ethylene dichloride produced is used to make polyvinyl chloride (PVC or vinyl). In fact, in the United States, about 40 percent of the chlorine produced goes to the manufacture of PVC. In addition to piping, tubing, flooring, siding, film, coatings, and many other products, PVC is also used in the production of prosthetic (artificial) limbs.

Another compound made using chlorine is propylene oxide (CH_3CHOCH_2). There is no chlorine in propylene oxide, but chlorine is used in the process by which the compound is made. Propylene oxide is used to make a group of plastics known as polyethers, the primary component of polyurethane foams. Polyethers are found in a wide range of materials, including car and boat bodies, bowling balls, fabrics for clothing, and rugs.

At one time, a large amount of chlorine was used to make a group of compounds known as chlorofluorocarbons (CFCs). CFCs are a family of chemical compounds containing **carbon**, fluorine, and chlorine. CFCs were once used in a wide variety of applications, such as air conditioning and refrigeration, aerosol spray products, and cleaning materials. They are now known to have serious environmental effects and have been banned from use in the United States and many other countries.

The reason for this ban is the damage caused by CFCs to Earth's ozone layer. Ozone (O_3) is a form of oxygen that filters out harmful radiation from the sun. When CFCs escape into the atmosphere, they attack and destroy ozone molecules. They reduce the protection against radiation provided by ozone.

Pesticides: DDT Chlorine has been used in making pesticides. A pesticide is a chemical used to kill pests. Pesticides have special names depending on the kind of pests they are designed to kill. Insecticides kill insects, rodenticides kill rodents (rats and mice), fungicides kill fungi, and nematicides kill worms.

Initially, people were unaware of the dangers posed by DDT. The pesticide was used on crops, in neighborhoods, and in other areas. Here, people are exposed to the pesticide while at the beach. LIBRARY OF CONGRESS.

Certain chlorine compounds have become very popular as pesticides. These compounds are called chlorinated hydrocarbons. They contain carbon, hydrogen, and chlorine.

Probably the most famous chlorinated hydrocarbon is dichlorodiphenyltrichloroethane, or DDT. First prepared in 1873, DDT was not used as a pesticide until World War II (1939–1945). Public health officials were at first delighted to learn that DDT kills disease-carrying insects very efficiently. There was great hope that DDT could be used to wipe out certain diseases in some parts of the world.

Farmers were also excited about DDT. They found it could kill many of the pests that attacked crops. By the end of the 1950s, many farmers were spraying huge amounts of DDT on their land to get rid of pests.

But problems began to appear. Many fish and birds in sprayed areas began to die or become deformed. Soon, these problems were traced to the use of DDT. The fish and birds ate insects that had been sprayed with DDT or drank water that contained DDT. It had a toxic effect on the fish and birds, just as it did on insects. Bird populations declined drastically as DDT caused eggs to be so thin-shelled that young birds did not survive.

Eventually, many governments began to ban the use of DDT, including the United States. It is still used in some nations, however, because the benefits of using DDT outweigh the harm it may cause. They believe that DDT can save many lives by killing insects that cause deadly diseases in humans, such as malaria. They know they can increase their food supplies by using DDT on crops as well. To lessen the risk of harming people, various governments follow strict procedures when using the pesticide.

DDT is not the only chlorinated hydrocarbon used as a pesticide. Other compounds in this class include dieldrin, aldrin, heptachlor, and chlordan. The use of these compounds has also been banned or restricted in the United States. The U.S. government has decided the harm they cause to the environment is more important than the benefits they provide to farmers and other users.

Health Effects

Chlorine gas is extremely toxic. In small doses, it irritates the nose and throat. A person exposed to chlorine may experience sneezing, running nose, and red eyes.

In larger doses, chlorine can be fatal. In fact, chlorine gas was used during World War I (1914–1918) by German soldiers as a biological weapon. Thousands of soldiers were killed or seriously wounded by breathing it. Those who survived gas attacks were often crippled for life. They were unable to breathe normally as a result of the damage to their throats and lungs.

On the other hand, chlorine compounds are essential to plants. They become sick or die without it. In plants, chlorine is regarded as a micronutrient, which is a substance needed in very small amounts to maintain good health. Leaves turn yellow and die when plants get too little chlorine from the soil.

Compounds of chlorine are important in maintaining good health in humans and other animals. The average human body contains about

95 grams (about 3.5 ounces) of chlorine. Hydrochloric acid (HCl) in the stomach, for example, helps in the digestion of foods. Sodium chloride (NaCl) and potassium chloride (KCl) play an important role in the way nerve messages are sent throughout the body. Because humans eat so much salt (NaCl), a lack of chlorine compounds is seldom a health problem.

Chromium

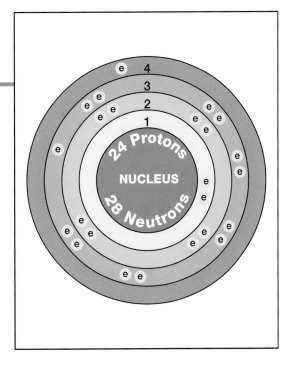

Key Facts

Symbol: Cr

Atomic Number: 24

Atomic Mass: 51.9961

Family: Group 6 (VIB); transition metal

Pronunciation: CRO-mee-um

Overview

Chromium is found in the center of the periodic table, a chart that shows how chemical elements are related to each other. Elements in Groups 3 through 12 are known as the transition metals. These elements all have similar physical and chemical properties. They have a bright, shiny surface and high melting points.

Chromium was discovered in 1797 by French chemist Louis-Nicolas Vauquelin (1763–1829). The element's name comes from the Greek word *chroma,* meaning "color," because chromium compounds are many different colors.

Much of the chromium produced today is used in alloys, including stainless steel. An alloy is made by melting and mixing two or more metals. The mixture has different properties than the individual metals. Chromium is also used to cover the surface of other metals. This technique protects the base metal and gives the surface a bright, shiny appearance at a low cost.

Discovery and Naming

Chromium was discovered in a mineral known as Siberian red lead. The mineral was first described in 1766 by German mineralogist Johann

WORDS TO KNOW

Abrasive: A powdery material used to grind or polish other materials.

Alloy: A mixture of two or more metals with properties different from those of the individual metals.

Catalyst: A substance used to speed up a chemical reaction without undergoing any change itself.

Electroplating: The process by which an electric current is passed through a water solution of a metallic compound.

Isotopes: Two or more forms of an element that differ from each other according to their mass number.

Periodic table: A chart that shows how the chemical elements are related to each other.

Radioactive isotope: An isotope that breaks apart and gives off some form of radiation.

Refractory: A material that can withstand very high temperatures and reflect heat away from itself.

Tracer: A radioactive isotope whose presence in a system can easily be detected.

Transition metal: An element in Groups 3 through 12 of the periodic table.

Gottlob Lehmann (1719–1767). Scientists were puzzled about what elements this new mineral contained. It had a form and a color not seen in other minerals. In some cases, it was found, as some said, "attached like little rubies to quartz."

Studies of Siberian red lead were difficult, however. It was mined at only one location in Germany and miners found it difficult to remove. Scientists had only small amounts of the mineral to study. They guessed that it contained **lead** as well as **arsenic**, **molybdenum**, or some other metal.

In 1797, Vauquelin began his own studies of Siberian red lead. He was convinced that the mineral contained a new element. None of the elements then known could account for his results. He reported finding "a new metal, possessing properties entirely unlike those of any other metal."

A year later, Vauquelin was able to isolate a small sample of the metal itself. He heated charcoal (nearly pure **carbon**) with a compound of chromium, chromium trioxide (Cr_2O_3). When the reaction was complete, he found tiny metallic needles of chromium metal:

$$2Cr_2O_3 + 3C \xrightarrow{\text{heat}} 3CO_2 + 4Cr$$

The name chromium was suggested by two French chemists, Antoine François de Fourcroy (1755–1809) and René-Just Haüy (1743–1822), because chromium forms so many different colored compounds. The colors range from purple and black to green, orange, and yellow.

Physical Properties

Chromium is a hard, steel-gray, shiny metal that breaks easily. It has a melting point of 3,450°F (1,900°C) and a boiling point of 4,788°F (2,642°C). The density is 7.1 grams per cubic centimeter. One important property is that chromium can be polished to a high shine.

Chemical Properties

Chromium is a fairly active metal. It does not react with water, but reacts with most acids. It combines with **oxygen** at room temperature to form chromium oxide (Cr_2O_3). Chromium oxide forms a thin layer on the surface of the metal, protecting it from further corrosion (rusting).

Occurrence in Nature

The abundance of chromium in Earth's crust is about 100 to 300 parts per million. It ranks about number 20 among the chemical elements in terms of their abundance in the earth.

Chromium does not occur as a free element. Today, nearly all chromium is produced from chromite, or chrome iron ore ($FeCr_2O_4$).

As of 2008, the leading producer of chromite ore is South Africa. Other important producers are Kazakhstan and India. According to the U.S. Geological Survey (USGS), most of the world's supply (95 percent) of chromium resources are in Kazakhstan and southern Africa. In the United States in 2008, one company was mining for chromite ore in Oregon.

Isotopes

There are four naturally occurring isotopes of chromium: chromium-50, chromium-52, chromium-53, and chromium-54. Isotopes are two or more forms of an element. Isotopes differ from each other according to their mass number. The number written to the right of the element's name is the mass number. The mass number represents the number of protons plus neutrons in the nucleus of an atom of the element. The number of protons determines the element, but the number of neutrons in the atom of any one element can vary. Each variation is an isotope.

Of the four naturally occurring isotopes, one (chromium-50) is radioactive. Seventeen radioactive isotopes of chromium have also been made in the laboratory. A radioactive isotope is one that breaks apart and gives off some form of radiation. Radioactive isotopes are produced when very small particles are fired at atoms. These particles stick in the atoms and make them radioactive.

One radioactive isotope of chromium is used in medical research, chromium-51. This isotope is used as a tracer in studies on blood. A tracer is a radioactive isotope whose presence in a system can be easily detected. The isotope is injected into the system at some point. Inside the system, the isotope gives off radiation. That radiation can be followed by means of detectors placed around the system.

A common use of chromium-51 is in studies of red blood cells. The isotope can be used to find out how many blood cells are present in a person's body. It can be used to measure how long the blood cells survive in the body. The isotope can also be used to study the flow of blood into and out of a fetus (an unborn child).

Extraction

The methods for producing chromium are similar to those used for other metals. One method is to heat chromium oxide (Cr_2O_3) with charcoal or **aluminum**. The charcoal (nearly pure carbon) or aluminum takes oxygen from the chromium oxide, leaving pure chromium metal. This method is similar to the one used by Vauquelin:

$$2Cr_2O_3 + 3C \longrightarrow_{heat} 3CO_2 + 4Cr$$

Chromium can also be obtained by passing an electric current through its some of its compounds:

$$2CrCl_3 \longrightarrow_{electric\ current} 2Cr + 3Cl_2$$

Sometimes chromite is converted directly to an alloy known as ferrochromium (or ferrochrome):

$$FeCr_2O_4 + C \longrightarrow_{heated} CO_2 + ferrochromium$$

Ferrochromium is an important chromium alloy. It is used to add chromium to steel. When steel is first made, it is a very hot, liquid material. To make chromium steel, ferrochromium is added to the hot liquid steel. There, the chromium dissolves into the hot steel. When the molten steel hardens, the chromium is trapped inside. It is now chromium steel.

Chromium comes from the Greek word for color. In these bottles, the different shades represent various colors of chromium compounds. © YOAV LEVY/PHOTOTAKE NYC.

Uses

Much of the chromium used in the United States goes into alloys. The addition of chromium makes the final product harder and more resistant to corrosion. Another significant use of chromium is in the production of stainless steel. The applications of stainless steel are almost endless. They include automobile and truck bodies, plating for boats and ships, construction parts for buildings and bridges, parts for chemical and petroleum equipment, electric cables, machine parts, eating and cooking utensils, and reinforcing materials in tires and other materials.

Two other major uses of chromium are electroplating and the manufacture of refractory bricks. Electroplating is the process by which an electric current is passed through a water solution of a metallic compound. The current causes the material to break down into two parts, as the following reaction shows:

$$2CrCl_3 \xrightarrow{\text{electric current}} 2Cr + 3Cl_2$$

The free chromium produced in this reaction is laid down in a thin layer on the surface of another metal, such as steel. The chromium protects the steel from corrosion and gives it a bright, shiny surface. Many kitchen appliances are "chrome-plated" this way.

Some chromium is also used to make refractory bricks. A refractory material can withstand very high temperatures by reflecting heat. Refractory materials are used to line high-temperature ovens.

Compounds

Chromium compounds have many different uses. Some include:

- chromic fluoride (CrF_3): printing, dyeing, and mothproofing woolen cloth
- chromic oxide (Cr_2O_3): a green pigment (coloring agent) in paint, asphalt roofing, and ceramic materials; refractory bricks; abrasive
- chromic sulfate ($Cr_2(SO_4)_3$): a green pigment in paint, ceramics, glazes, varnishes, and inks; chrome plating
- chromium boride (CrB): refractory; high-temperature electrical conductor
- chromium dioxide (CrO_2): covering for magnetic tapes ("chromium" tapes)
- chromium hexacarbonyl ($Cr(CO)_6$): catalyst; gasoline additive

Health Effects

Chromium is an element with two faces, as far as health effects are concerned. Small amounts of chromium are essential for the health of plants and animals. In humans, a chromium deficiency leads to diabetes-like symptoms. Diabetes is a disease that develops when the body does not use sugar properly. Chromium seems to play a role in helping the body use sugar.

In larger amounts, chromium is harmful. Some compounds are especially dangerous, causing a rash or sores if spilled on the skin. They can also cause sores in the mouth and throat if inhaled. If swallowed, some chromium compounds can seriously damage the throat, stomach, intestines, kidneys, and circulatory (blood) system. Scientists believe exposure to some chromium compounds on a long-term basis causes cancer. As a result, the U.S. Environmental Protection Agency (EPA) has established rules about the amount of chromium to which workers can be exposed.

Cobalt

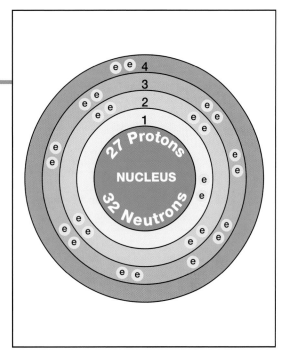

Overview

Humans have been using compounds of cobalt since at least 1400 BCE. The compounds were used to color glass and glazes blue. In 1735, Swedish chemist Georg Brandt (1694–1768) analyzed a dark blue pigment found in copper ore. Brandt demonstrated that the pigment contained a new element, later named cobalt.

Cobalt is a transition metal, one of the elements found in Rows 4 through 7 between Groups 2 and 13 in the periodic table. The periodic table is a chart that shows how chemical elements are related to each other. Cobalt is located between **iron** and **nickel** and shares many chemical and physical properties with these two elements.

The United States has to import all the cobalt it uses. One of the most important applications of cobalt is in the production of superalloys. These superalloys consist primarily of iron, cobalt, or nickel, with small amounts of other metals, such as **chromium**, **tungsten**, **aluminum**, and **titanium**. Superalloys are resistant to corrosion (rusting) and retain their properties at high temperatures. Superalloys are used in jet engine parts and gas turbines.

Key Facts

Symbol: Co

Atomic Number: 27

Atomic Mass: 58.933195

Family: Group 9 (VIIIB); transition metal

Pronunciation: CO-balt

WORDS TO KNOW

Alloy: A mixture of two or more metals with properties different from those of the individual metals.

Catalyst: A substance used to speed up a chemical reaction without undergoing any change itself.

Isotopes: Two or more forms of an element that differ from each other according to their mass number.

Magnetic field: The space around an electric current or a magnet in which a magnetic force can be observed.

Malleable: Capable of being hammered into thin sheets.

Periodic table: A chart that shows how chemical elements are related to each other.

Radioactive isotope: An isotope that breaks apart and gives off some form of radiation.

Superalloys: Consist primarily of iron, cobalt, or nickel, with small amounts of other metals that are resistant to corrosion (rusting) and retain their properties at high temperatures.

Trace mineral: An element needed by plants and animals in minute amounts.

Transition metal: An element in Groups 3 through 12 of the periodic table.

Discovery and Naming

Cobalt dyes have been used for centuries. Craftsmen used materials from the earth to color glass, pottery, glazes, and other materials. Cobalt minerals were especially prized for their rich blue color.

The word cobalt may date back to the end of the 15th century. In German, the word *Kobold* means "goblin" or "evil spirit." The term was used by miners to describe a mineral that was very difficult to mine and was damaging to their health. When the mineral was heated, it gave off an offensive gas that caused illness. The gas that affected the miners was **arsenic** trioxide (As_4O_6), which often occurs with cobalt in nature.

At first, chemists were skeptical about Brandt's claims of a new element, but he continued his research on the mineral. He showed that its compounds were a much deeper blue than copper compounds. (Copper and cobalt compounds had long been confused with each other.) Eventually, Brandt was given credit for the discovery of the element. The name chosen was a version of the original German term, Kobold.

Physical Properties

Cobalt is a hard, gray metal that looks much like iron and nickel. It is ductile, but only moderately malleable. Ductile means capable of being

drawn into thin wires. Malleable means capable of being hammered into thin sheets.

Cobalt is one of only three naturally occurring magnetic metals. The other two are iron and nickel. The magnetic properties of cobalt are even more obvious in alloys. An alloy is made by melting and mixing two or more metals. The mixture has properties different from those of the individual metals.

The melting point of cobalt metal is 2,719°F (1,493°C), and the boiling point is about 5,600°F (3,100°C). The density is 8.9 grams per cubic centimeter.

Chemical Properties

Cobalt is a moderately reactive element. It combines slowly with **oxygen** in the air, but does not catch fire and burn unless it is in a powder form. It reacts with most acids to produce **hydrogen** gas. It does not react with water at room temperatures.

Occurrence in Nature

Cobalt is a relatively abundant element at about 10 to 30 parts per million. This places it in the upper third of elements according to their abundance in Earth's crust.

The most common ores of cobalt are cobaltite, smaltite, chloranthite, and linnaeite. As of 2008, the major suppliers of cobalt in the world are Congo (Kinshasa), Canada, Zambia, Australia, Russia, Cuba, and China. No cobalt was mined or refined in the United States, although significant resources exist in the country. According to the U.S. Geological Survey (USGS), states with cobalt resources include: Alaska, California, Idaho, Minnesota, Missouri, Montana, and Oregon.

Isotopes

There is only one naturally occurring isotope of cobalt: cobalt-59. Isotopes are two or more forms of an element. Isotopes differ from each other according to their mass number. The number written to the right of the element's name is the mass number. The mass number represents the number of protons plus neutrons in the nucleus of an atom of the element. The number of protons determines the element, but the number of neutrons in the atom of any one element can vary. Each variation is an isotope.

Twenty-six radioactive isotopes of cobalt are known also, for which half lives are available. A radioactive isotope is one that breaks apart and gives off some form of radiation. Radioactive isotopes are produced when very small particles are fired at atoms. These particles stick in the atoms and make them radioactive.

Cobalt-60 One of the most widely used of all radioactive isotopes is cobalt-60. In medicine, it is used to find and treat diseases. For example, it is used in a test known as the Schilling test. This test is a method for determining whether a person's body is making and using vitamin B_{12} properly. Two other isotopes of cobalt, cobalt-57 and cobalt-58, are used for the same purpose.

Cobalt-60 is also used to treat cancer. The radiation given off by the isotope kills cancer cells. The isotope has been used for more than 50 years to treat various forms of cancer.

A growing use of cobalt-60 is in food irradiation. Food irradiation is a method for preserving food. The food is exposed to radiation from cobalt-60. That radiation kills bacteria and other organisms that cause disease and spoilage. The food can be stored longer without going bad after being irradiated.

There is some controversy about the use of irradiation as a way of preserving food. Some people worry that harmful compounds will be produced during irradiation. So far, no proof has been found that irradiation is a dangerous method of food preservation.

Cobalt-60 is also used in industrial applications. The radiation it gives off acts like X rays from an X-ray machine. It can penetrate metals. The X-ray pattern produced by radiating a material tells about its strength, composition, and other properties.

Extraction

Cobalt is obtained by heating its ores to produce cobalt oxide (Co_2O_3). That compound is then heated with aluminum to free the pure metal:

$$2Al + Co_2O_3 \rightarrow Al_2O_3 + 2Co$$

In a second method, cobalt oxide is first converted to cobalt chloride ($CoCl_3$). An electric current is then passed through molten (melted) cobalt chloride to obtain the free element:

$$2CoCl_3 \xrightarrow{\text{electric current}} 2Co + 3Cl_2$$

Uses

About 46 percent of cobalt used in the United States is used to make alloys, mostly superalloys. These superalloys are used in situations where metals are placed under extreme stress, often at high temperatures. A gas turbine, a device for making electricity, is a good example. It looks a bit like a large airplane propeller, with many blades. A hot, high speed gas pushes against the turbine blades, making them spin very fast. The motion generates electricity. Cobalt superalloys hold up to the high temperature stress produced in the machine.

Cobalt is also used in making magnetic alloys. These alloys are used to make devices that must hold a magnetic field, such as electric motors and generators. A magnetic field is the space around an electric current or a magnet in which a magnetic force can be observed. Another application of cobalt alloys is in the production of cemented carbides. In metallurgy, cementation is the process by which one metal is covered with a fine coating of a second metal. Cementation is used to make very hard, strong alloys, such as those used in drilling tools and dies.

The primary use of cobalt worldwide is in the production of rechargeable battery electrodes.

Compounds

Cobalt compounds are widely used to make coloring materials. The following compounds are used to color glass, glazes, cosmetics, paints, rubber, inks, and pottery: cobalt oxide, or cobalt black (Co_2O_3); cobalt potassium nitrite, or cobalt yellow ($CoK_3(NO_2)_6$); cobalt aluminate, or cobalt blue ($Co(AlO_2)_2$); and cobalt ammonium phosphate, or cobalt violet ($CoNH_4PO_4$).

Another important use of cobalt compounds is as catalysts. A catalyst is a substance used to speed up a chemical reaction. The catalyst does not undergo any change itself during the reaction. Cobalt molybdate ($CoMoO_4$) is used in the petroleum industry to convert crude oil to gasoline and other petroleum products. It is also used to remove **sulfur** from crude oil.

Health Effects

Cobalt is a trace mineral in the human body. A trace mineral is an element needed by plants and animals in minute amounts. The absence of a trace mineral in the diet often leads to health problems. In animals, cobalt is used to make certain essential enzymes. An enzyme is a catalyst in a living organism. It speeds up the rate at which certain changes take place in the body. Enzymes are essential in order for living cells to function properly. Cobalt is needed for the production of vitamin B_{12}. Vitamin B_{12} is necessary to ensure that an adequate number of red blood cells is produced in the body.

A lack of cobalt in the soil can cause health problems, too. For example, sheep in Australia are subject to a disease known as Coast disease, due to a deficiency of cobalt in the soil.

An excess of cobalt can also cause health problems. For example, people who work with the metal may inhale its dust or get the dust on their skin. Cobalt dust can cause vomiting, diarrhea, or breathing problems. On the skin, it can cause rashes and irritation.

Copper

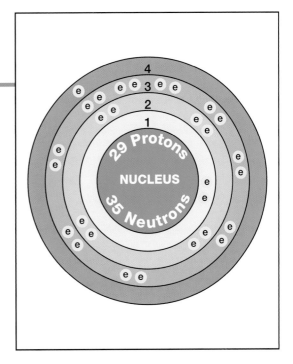

Overview

Copper was one of the earliest elements known to humans. At one time, it could be found lying on the ground in its native, or uncombined, state. Copper's distinctive red color made it easy to identify. Early humans used copper for many purposes, including jewelry, tools, and weapons.

Copper is a transition metal, one of the elements found in Rows 4 through 7 between Groups 2 and 13 in the periodic table. The periodic table is a chart that shows how chemical elements are related to each other.

Copper and its compounds have many important uses in modern society. For example, copper wiring is used in electrical equipment. Copper is also used to make many alloys. An alloy is made by melting and mixing two or more metals. The mixture has properties different from those of the individual metals. The most familiar alloys of copper are probably brass and bronze. Many compounds of copper are commercially important, too. They are used as coloring agents in paints, ceramics, inks, varnishes, and enamels.

Discovery and Naming

The oldest known objects made of copper are beads found in northern Iraq, which date to about 9000 BCE. Tools for working with copper,

Key Facts

Symbol: Cu

Atomic Number: 29

Atomic Mass: 63.546

Family: Group 11 (IB); transition metal

Pronunciation: COP-per

WORDS TO KNOW

Alkali: A chemical with properties opposite those of an acid.

Alloy: A mixture of two or more metals with properties different from those of the individual metals.

Ductile: Capable of being drawn into thin wires.

Electrolysis: The process by which an electrical current is used to cause a chemical change, usually the breakdown of some substance.

Enzyme: A substance that stimulates certain chemical reactions in the body.

Isotopes: Two or more forms of an element that differ from each other according to their mass number.

Periodic table: A chart that shows how the chemical elements are related to each other.

Radioactive isotope: An isotope that breaks apart and gives off some form of radiation.

Slurry: A soup-like mixture of crushed ore and water.

Transition metal: An element in Groups 3 through 12 of the periodic table.

made in about 5000 BCE, have also been found. In the New World, Native Americans used copper objects as early as 2000 BCE.

The symbol for copper, Cu, comes from the Latin word *cuprum*. Cuprum is the ancient name of the island of Cyprus, which is located in the Mediterranean Sea near Turkey. The Romans obtained much of their copper from Cyprus.

Bronze was one of the first alloys produced. It consists primarily of copper and **tin**. The two metals can be melted together rather easily. Humans discovered methods for making the alloy as early as 4000 BCE. Bronze was used for a great variety of tools, weapons, jewelry, and other objects. It was such an important metal that the period from 3500 to 1000 BCE is now known as the Bronze Age. The Iron Age followed the Bronze Age when **iron** began to replace bronze in tools and weapons.

Physical Properties

An important physical property of copper is its color. In fact, people often refer to anything with a reddish-brown tint as being copper colored.

Copper metal is fairly soft and ductile. Ductile means capable of being drawn into wires. Both heat and electricity pass through copper very easily. The high electrical conductivity makes it ideal for many electrical purposes.

Copper has a melting point of 1,982°F (1,083°C) and a boiling point of 4,703°F (2,595°C). Its density is 8.96 grams per cubic centimeter.

Chemical Properties

Copper is a moderately active metal. It dissolves in most acids and in alkalis. An alkali is a chemical with properties opposite those of an acid. Sodium hydroxide, commonly found in bleach and drain cleaners, is an example of an alkali. An important chemical property of copper is the way it reacts with **oxygen**. In moist air, it combines with water and carbon dioxide. The product of this reaction is called hydrated copper carbonate ($Cu_2(OH)_2CO_3$), which changes copper's reddish-brown color to a beautiful greenish color, called a patina. Copper roofs eventually develop this color.

Statue of Liberty Perhaps the most famous and dramatic example of this phenomenon is the Statue of Liberty on Ellis Island near New York City. The Statue, or Lady Liberty as it is often called, was a gift to the United States from France. It was dedicated on October 28, 1886. It symbolizes political freedom and democracy.

The Statue of Liberty. IMAGE COPYRIGHT 2009, DONALD R. SWARTZ. USED UNDER LICENSE FROM SHUTTERSTOCK.COM.

The Statue of Liberty is covered with copper plates. When it was new, Lady Liberty was copper in color. Over time, the plates slowly turned green. The statue was given a thorough cleaning for its 100th birthday party on July 4, 1986. But the color remained green. It would take a lot of elbow grease to return the Lady to her original copper color.

Occurrence in Nature

The abundance of copper in Earth's crust is estimated to be about 70 parts per million. It ranks in the upper quarter among elements present in Earth's crust. Small amounts (about 1 part per billion) also occur in seawater.

At one time, it was not unusual to find copper metal lying on the ground. However, native copper can still be found only rarely today, always in areas where humans seldom travel.

Today, essentially all copper is obtained from minerals such as azurite, or basic copper carbonate ($Cu_2(OH)_2CO_3$); chalcocite, or copper glance or copper sulfide (Cu_2S); chalcopyrite, or copper pyrites or copper iron sulfide ($CuFeS_2$); cuprite, or copper oxide (Cu_2O); and malachite, or basic copper carbonate ($Cu_2(OH)_2CO_3$).

Copper is mined in more than 50 nations, from Albania and Argentina to Zambia and Zimbabwe. As of 2008, the leading producers are Chile, the United States, Peru, and China. The next largest producers are Australia, Russia, Indonesia, and Canada. According to the U.S. Geological Survey (USGS), about 99 percent of copper mined in the United States comes from Arizona, Utah, New Mexico, Nevada, and Montana.

Isotopes

There are two naturally occurring isotopes of copper: copper-63 and copper-65. Isotopes are two or more forms of an element. Isotopes differ from each other according to their mass number. The number written to the right of the element's name is the mass number. The mass number represents the number of protons plus neutrons in the nucleus of an atom of the element. The number of protons determines the element, but the number of neutrons in the atom of any one element can vary. Each variation is an isotope.

Twenty-four radioactive isotopes of copper are known also. A radioactive isotope is one that breaks apart and gives off some form of radiation. Radioactive isotopes are produced when very small particles are fired at atoms. These particles stick in the atoms and make them radioactive.

Two radioactive isotopes of copper are used in medicine. One is copper-64. This isotope is used to study brain function and to detect Wilson's disease. This disease is the inability to eliminate copper from one's body. The second isotope is copper-67. This isotope can be used to treat cancer. The isotope is injected into the body. It then goes to cells that have become cancerous. In these cells, the isotope gives off radiation that can kill the cancerous cells.

Extraction

Converting copper ore to copper metal often involves many steps. First, the ore is crushed into small pieces. Then the crushed pieces are mixed with water to form a slurry, a soup-like mixture of crushed ore and water.

The slurry is spun around in large vats with steel balls to crush the ore to an even finer powder.

Next, blasts of air are passed through the slurry. Impure copper rises to the top of the mixture and unwanted earthy materials sink to the bottom. The copper mixture is skimmed off the top of the slurry and dissolved in sulfuric acid (H_2SO_4).

Bars of **iron** are added to the copper/sulfuric acid mixture. Iron is a more active metal than copper. It replaces the copper from the sulfuric acid solution. Copper deposits on the surface of the iron bar where it is easily scraped off.

The copper is still not pure enough for most purposes. The most common method for copper purification relies on electrolysis. Electrolysis is the process by which an electrical current is used to cause a chemical change, usually the breakdown of some substance. The copper is dissolved in sulfuric acid again and an electric current is passed through the solution. Pure copper metal is deposited on one of the metal electrodes. By repeating this process, copper of 99.9 percent purity can be made.

Other methods are also used to remove copper from its ores. The method chosen depends on the kind of ore used.

Uses

One of the most important applications of copper metal is electrical wiring. Many electrical devices rely on copper wiring because copper metal is highly conductive and relatively inexpensive. These devices include electric clocks, stoves, portable CD and DVD players, and transmission wires that carry electricity. A large skyscraper contains miles of copper wiring for all its electrical needs. Older telephone lines are thick bundles of copper wires. And computers contain circuit boards imprinted with minute copper pathways.

Alloys of copper, such as bronze and brass, are also used in construction. These alloys are used in roofs, heating and plumbing systems, and the skeleton of the building itself.

A number of copper alloys have been developed for special purposes. For example, gun metal is an alloy used to make guns. It contains about 90 percent copper and 10 percent tin. Monel metal is an alloy of nickel and copper that is resistant to corrosion (rusting). Coinage metal is a copper alloy from which U.S. coins are made.

The Value of a Penny When the penny was first introduced in the United States in 1787, it consisted of pure copper. The first Lincoln penny, released in 1909, was 95 percent copper. Depending on the year, the other 5 percent was either all zinc or a combination of zinc and tin (bronze). In 1943—during World War II (1939–1945)—the penny consisted of zinc-plated steel. This penny was a failure. The steel was magnetic (so it got stuck in vending machines), the zinc corroded easily, and the public often confused it with a dime.

By the 1980s, copper had become more valuable than the one cent that the penny was worth. So in 1982, the U.S. mint switched the penny's core to an inexpensive zinc coated with copper. The rest of U.S. pocket change—dimes, nickels, and quarters—have a core of coinage metal with a thin coating of a silvery metal.

Compounds

A number of copper compounds are used as pesticides, chemicals that kill insects and rodents like rats and mice:

- basic copper acetate ($Cu_2O(C_2H_3O_2)_2$): insecticide (kills insects) and fungicide (kills fungi)
- copper chromate ($CuCrO_4 \cdot 2CuO$): fungicide for the treatment of seeds
- copper fluorosilicate ($CuSiF_6$): grapevine fungicide

- copper methane arsenate ($CuCH_3AsO_3$): algicide (kills algae)
- copper-8-quinolinolate ($Cu(C_9H_6ON)_2$): protects fabric from mildew
- copper oxalate (CuC_2O_4): seed coating to repel rats
- copper oxychloride ($3CuO \cdot CuCl_2$): grapevine fungicide
- tribasic copper sulfate ($CuSO_4 \cdot 3Cu(OH)_2$): fungicide, used as a spray or dust on crops

Other copper compounds are found in battery fluid; fabric dye; fire retardants; food additives for farm animals; fireworks (bright emerald color); manufacture of ceramics and enamels; photographic film; pigments (coloring agents) in paints, metal preservatives, and marine paints; water purification; and wood preservatives.

Turquoise and malachite are semi-precious gemstones made up of copper compounds. Turquoise ranges in color from green to blue.

Blue-Blooded Creatures In humans, the blood that comes from the lungs to the cells is bright red. The red color is caused by oxyhemoglobin (the compound hemoglobin combined with oxygen). Hemoglobin carries oxygen through the blood and is red because of the iron it carries. Compounds of iron are often red or reddish-brown. Blood returning from cells to the lungs (which flows through the veins) is purplish-red because the hemoglobin has lost its oxygen.

Some animals, however, do not have hemoglobin to carry oxygen through the blood. For example, crustaceans (shellfish like lobsters, shrimps, and crabs) use a compound called hemocyanin. Hemocyanin is similar to hemoglobin but contains copper instead of iron. Many copper compounds, including hemocyanin, are blue. Therefore, the blood of a crustacean is blue, not red.

Health Effects

Copper is an essential micronutrient for both plants and animals. A micronutrient is an element needed in minute amounts to maintain good health in an organism. A healthy human has no more than about 2 milligrams of copper for every 2.2 pounds (1 kilogram) of body weight.

Copper is critical to the production of some enzymes. An enzyme is a substance that stimulates certain chemical reactions in the body. Without enzymes, the reactions would be too slow. Copper enzymes function in the production of blood vessels, tendons, bones, and nerves. Animals seldom become ill from a lack of copper, but copper-deficiency disorders (problems because of lack of copper) can occur with animals who live on land that lacks copper.

Crustaceans have blue blood because of a copper compound called hemocyanin. IMAGE COPYRIGHT 2009, DMITRIJS MIHEJEVS. USED UNDER LICENSE FROM SHUTTERSTOCK.COM.

Large amounts of copper in the human body are usually not a problem either. One exception is the condition known as Wilson's disease. Some people are born without the ability to eliminate copper from their bodies. The amount of copper they retain increases. The copper level can become so great it begins to affect a person's brain, liver, or kidneys. Mental illness and death can result. Fortunately, this problem can be treated. The person is given a chemical that combines with the copper. The copper's damaging effects on the body are reduced or eliminated.

Curium

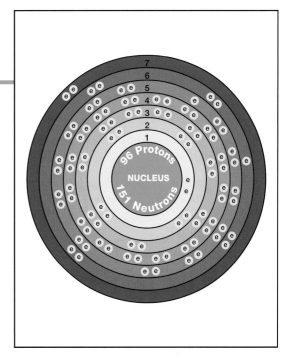

Overview

Curium is called a transuranium element because it follows **uranium** in the periodic table. The periodic table is a chart that shows how chemical elements are related to each other. Uranium has an atomic number of 92, so any element with a higher atomic number is a transuranium element.

Curium was discovered in 1944 by Glenn Seaborg (1912–1999), Ralph A. James, and Albert Ghiorso (1915–). These researchers, from the University of California at Berkeley, were working at the Metallurgical Research Laboratory (MRL) at the University of Chicago where research on the first atomic bomb was being conducted.

Discovery and Naming

Curium was first produced in a particle accelerator at the MRL. A particle accelerator is also called an atom smasher. It is used to accelerate small particles, such as protons, to move at very high speeds. The particles approach the speed of light, 186,000 miles per second (300,000 kilometers per second), and collide with target elements, such as **gold**, **copper**, or **tin**. The targets break apart or combine with a particle to form new elements and other particles.

Key Facts

Symbol: Cm

Atomic Number: 96

Atomic Mass: [247]

Family: Actinoid; transuranium element

Pronunciation: CURE-ee-um

The first samples of curium were so small they could be detected only by the radiation they gave off. In 1947, the first significant sample of the element was produced. It weighed about 30 milligrams, or the equivalent of about one-thousandth of an ounce. The element was named for Polish-French physicist Marie Curie (1867–1934) and her husband, French physicist Pierre Curie (1859–1906). The Curies carried out some of the earliest research on radioactive elements.

Physical Properties

Curium is a silvery-white metal with a melting point of about 2,444°F (1,340°C) and a density of 13.5 grams per cubic centimeter, about 13 times the density of water.

Chemical Properties

The chemical properties of curium are similar to those of the rare earth elements. The rare earth elements are the elements in the periodic table with atomic numbers 57 through 71. Curium is not very reactive at room temperature, but does combine with oxygen when heated to form curium oxide (Cm_2O_3). The element also reacts with the halogens to form curium fluoride (CmF_4), curium chloride ($CmCl_3$), curium bromide ($CmBr_3$), and curium iodide (CmI_3). A number of other curium compounds have also been made.

Curium was named after Polish-French physicists Marie and Pierre Curie, who conducted research on radioactive elements.
AP IMAGES.

Occurrence in Nature

Very small amounts of curium are thought to occur in Earth's surface with deposits of uranium. The curium is formed when uranium breaks down and forms new elements. The amounts that exist, if they do, are too small to have been discovered so far.

Isotopes

All 21 known isotopes of curium are radioactive. Isotopes are two or more forms of an element. Isotopes differ from each other according to their mass number. The number written to the right of the element's name is the mass number. The mass number represents the number of protons plus neutrons in the nucleus of an atom of the element. The number of protons determines the element, but the number of neutrons in the atom of any one element can vary. Each variation is an isotope. A radioactive isotope is one that breaks apart and gives off some form of radiation.

The curium isotope with the longest half life is curium-247. Its half life is 15.6 million years. The half life of a radioactive element is the time it takes for half of a sample of the element to break down. After about 15.6 million years, only 0.5 grams of the isotope would remain from a one-gram sample produced today. The other 0.5 gram would have changed into another element.

Extraction

Large quantities of curium are now easily made in nuclear reactors. A nuclear reactor is a device in which neutrons split atoms to release energy for electricity production.

Uses

Curium is sometimes used to analyze materials taken from mines and as a portable source of electrical power. It gives off a large amount of energy that can be used to generate electricity for space vehicles.

A relatively recent use of curium was in the *Mars Pathfinder* that was sent to Mars in 1997 to study that planet's surface. Some of the equipment on the spacecraft was powered by a curium battery.

Compounds

A number of compounds of curium have been produced, including two forms of curium oxide (Cm_2O_3 and CmO_2), two forms of curium fluoride (CmF_3 and CmF_4), curium chloride ($CmCl_3$), curium bromide ($CmBr_3$), and curium hydroxide ($Cm(OH)_3$).

Health Effects

Curium is an extremely hazardous substance. If taken into the body, it tends to concentrate in the bones, where the radiation it gives off kills or damages cells and can cause cancer.

Dysprosium

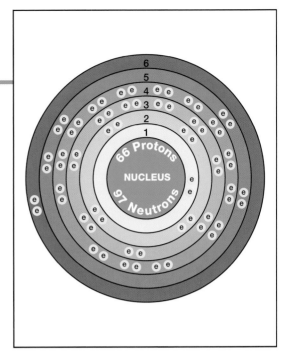

Overview

Dysprosium is one of 15 rare earth elements in Row 6 of the periodic table. The name rare earth is misleading because the elements in this group are not especially uncommon. However, they often occur together in the earth and were once difficult to separate from each other. A better name for the rare earth elements is lanthanoids. This name comes from the element **lanthanum**, which is sometimes considered part of the lanthanoids group in the periodic table. The periodic table is a chart that shows how chemical elements are related to one another.

Dysprosium was discovered in 1886, but was not commercially available until after 1950. The reason for the long delay was that methods for separating dysprosium from other lanthanoids had not been developed. Dysprosium has few applications but is used in some hybrid cars.

Discovery and Naming

In 1787, Carl Axel Arrhenius (1757–1824), a Swedish army officer and amateur mineralogist, found a rock in a mine at Ytterby, a region near Stockholm. He named the rock ytterite. A few years later, the rock was analyzed by Johan Gadolin (1760–1852), a professor of chemistry at the

Key Facts

Symbol: Dy

Atomic Number: 66

Atomic Mass: 162.500

Family: Lanthanoid (rare earth metal)

Pronunciation: dis-PRO-zee-um

WORDS TO KNOW

Earth: In mineralogy, a naturally occurring form of an element, often an oxide of the element.

Isotopes: Two or more forms of an element that differ from each other according to their mass number.

Lanthanoids (rare earth elements): The elements in the periodic table with atomic numbers 57 through 71.

Periodic table: A chart that shows how chemical elements are related to each other.

University of Åbo in Finland, who found that the rock contained a new kind of "earth," which a colleague called yttria.

The term "earth" in mineralogy refers to a naturally occurring form of an element, usually an oxide. For example, one kind of earth is magnesia, a term that refers to **magnesium** oxide. Magnesium oxide is one form in which the element magnesium occurs naturally in the earth.

The discovery by Arrhenius and Gadolin initiated a long series of experiments on yttria. These experiments produced puzzling results for two reasons. First, it turned out that yttria is actually a mixture of many similar elements. Second, the equipment that chemists had to work with was still very primitive. They had serious difficulties separating these elements from each other.

Over a period of more than a century, chemists argued about the composition of yttria. Eventually, chemists agreed that yttria is actually a combination of nine different elements that had not been seen before. One of those elements is dysprosium. Dysprosium was finally proved to be a new element in 1886 by French chemist Paul-émile Lecoq de Bois-baudran (1838–1912). The name chosen for this new element comes from the Greek word meaning "difficult to obtain."

Physical Properties

Dysprosium has a metallic appearance with a shiny silver luster. The metal is so soft it is easily cut with a knife. It has a melting point of 2,565°F (1,407°C) and a boiling point of about 4,200°F (about 2,300°C). Dysprosium's density is 8.54 grams per cubic centimeter, about eight times that of water.

Chemical Properties

Dysprosium is relatively unreactive at room temperatures. It does not oxidize very rapidly when exposed to the air. It does react with both dilute and concentrated acids, however. For example, it reacts with hydrochloric acid to form dysprosium trichloride.

$$2Dy + 6HCl \rightarrow 2DyCl_3 + 3H_2$$

Occurrence in Nature

More than 100 minerals are known to contain one or more of the rare lanthanoids. Only two of these minerals, monazite and bastnasite, are commercially important. These minerals occur in North and South Carolina, Idaho, Colorado, and Montana in the United States, and in countries such as Australia, Brazil, China, and India, among others.

Experts estimate that no more than about 8.5 parts per million of dysprosium occur in Earth's crust. That makes the element more common than better known elements such as **bromine**, **tin**, and **arsenic**. Studies of stony meteorites have found about 0.3 parts per million of dysprosium.

Isotopes

Seven naturally occurring isotopes of dysprosium are known. Isotopes are two or more forms of an element. Isotopes differ from each other according to their mass number. The number written to the right of the element's name is the mass number. The mass number represents the number of protons plus neutrons in the nucleus of an atom of the element. The number of protons determines the element, but the number of neutrons in the atom of any one element can vary. Each variation is an isotope. The four most abundant isotopes of dysprosium are dysprosium-161, dysprosium-162, dysprosium-163, and dysprosium-164.

Twenty-eight radioactive isotopes of dysprosium have been made. A radioactive isotope is one that is unstable, gives off radiation, and breaks down to form a new isotope. Of these isotopes, only one, dysprosium-166, has much commercial importance. It is used to treat certain types of cancer, to relieve pain due to cancer, and in the treatment of joint problems.

The radioactive isotope dysprosium-165 is also being studied for some potential applications in medicine. Radiation with dysprosium-165 has proved to be more effective in treating damaged joints than traditional surgery.

Extraction

The dysprosium in monazite and bastnasite is first converted to dysprosium trifluoride (DyF_3). The compound then reacts with calcium metal to obtain pure dysprosium:

$$2DyF_3 + 3Ca \rightarrow 3CaF_2 + 2Dy$$

Uses

Dysprosium has a tendency to soak up neutrons, which are tiny particles that occur in atoms and are produced in nuclear reactions. Metal rods (control rods) containing dysprosium are used in nuclear reactors to control the rate at which neutrons are available.

Dysprosium is also used to make alloys for various electrical and electronic devices. An alloy is made by melting and mixing two or more metals. The mixture has properties different than any of the elements. Some dysprosium alloys have very good magnetic properties that make them useful in CD players. The element is also used in some hybrid cars.

The magnetic properties of dysprosium alloys make them useful in CD players. IMAGE COPYRIGHT 2009, SILVER-JOHN. USED UNDER LICENSE FROM SHUTTERSTOCK.COM.

Compounds

Like the element itself, some compounds of dysprosium are used in nuclear reactors and the manufacture of electrical and electronic equipment.

Health Effects

Very little is known about the health effects of dysprosium.

Einsteinium

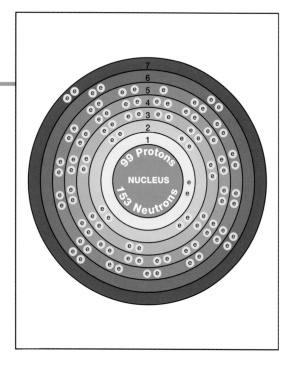

Overview

Einsteinium is a member of the actinoid family. The actinoid elements are found in Row 7 of the periodic table, a chart that shows how chemical elements are related to each other. The actinoids fall between **radium** (element number 88) and **rutherfordium** (element number 104). They are usually listed in a separate row at the very bottom of the periodic table.

Einsteinium is also a transuranium element. Transuranium elements are those beyond **uranium** on the periodic table. Uranium has an atomic number of 92, so elements with larger atomic numbers are transuranium elements.

Discovery and Naming

Einsteinium was discovered by a research team from the University of California at Berkeley. The team was led by Albert Ghiorso (1915–). The element was discovered in the "ashes" after the first hydrogen bomb test in November 1952 at Eniwetok Atoll, Marshall Islands, in the Pacific Ocean. The discovery was a remarkable accomplishment because no more than a hundred millionth of a gram of the element was present. It was detected because of the characteristic radiation it produced.

Key Facts

Symbol: Es

Atomic Number: 99

Atomic Mass: [252]

Family: Actinoid; transuranium element

Pronunciation: ein-STY-nee-um

Element number 99 was named after German-American physicist Albert Einstein (1879–1955). Some people regard Einstein as the greatest scientist who ever lived.

The element einsteinium is named after Albert Einstein.
LIBRARY OF CONGRESS.

Physical and Chemical Properties

Very little information about the physical and chemical properties of einsteinium is available. Its melting point has been found to be 1,580°F (860°C).

Occurrence in Nature

Einsteinium does not occur naturally in Earth's crust.

Isotopes

All 19 isotopes of einsteinium are radioactive. The most stable is einsteinium-254. Its half life is 275.7 days. Isotopes are two or more forms of an element. Isotopes differ from each other according to their mass number. The number written to the right of the element's name is the mass number. The mass number represents the number of protons plus neutrons in the nucleus of an atom of the element. The number of protons determines the element, but the number of neutrons in the atom of any one element can vary. Each variation is an isotope. A radioactive isotope is one that breaks apart and gives off some form of radiation.

The half life of a radioactive element is the time it takes for half of a sample of the element to break down. For example, suppose that scientists made 10 grams of einsteinium-254. About eight months later (275.7 days later), only 5 grams of the element would be left. After another eight months (275.7 days more), only half of that amount (2.5 grams) would remain.

Extraction

Einsteinium is not extracted from Earth's crust.

Uses

Einsteinium is sometimes used for research purposes, but it has no practical applications.

Compounds

There are no commercially important compounds of einsteinium.

Health Effects

Scientists know too little about einsteinium to be aware of its health effects. As a radioactive element, however, it does pose a threat to human health.

Erbium

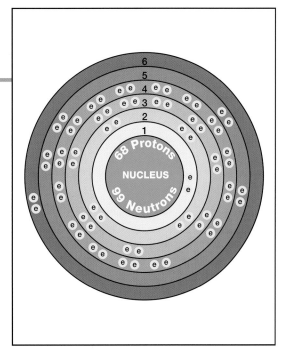

Overview

Erbium is one of 15 rare earth elements with atomic numbers 57 through 71 in Row 6 of the periodic table. The periodic table is a chart that shows how chemical elements are related to each other. Rare earth is a misleading category because these elements are not especially rare in Earth's crust. Rare earth metals were rarely used because the elements were once difficult to separate from each other.

Today, the rare earth elements can be separated easily, and they can be bought at a reasonable cost. Two common uses of erbium today are in lasers and special kinds of optical fibers. Optical fibers are glass-like materials used to carry telephone messages.

The rare earth elements are called the lanthanoids, a name that comes from **lanthanum** (element 57).

Discovery and Naming

The discovery of the lanthanoids began outside the small town of Ytterby, Sweden, in 1787. A Swedish army officer named Carl Axel Arrhenius (1757–1824) found an unusual kind of black mineral in a rock quarry. That mineral was later given the name gadolinite.

Key Facts

Symbol: Er

Atomic Number: 68

Atomic Mass: 167.259

Family: Lanthanoid (rare earth metal)

Pronunciation: ER-bee-um

WORDS TO KNOW

Alloy: A mixture of two or more metals with properties different from those of the individual metals.

Isotopes: Two or more forms of an element that differ from each other according to their mass number.

Lanthanoids (rare earth elements): The elements in the periodic table with atomic numbers 57 through 71.

Laser: A device for making very intense light of one very specific color that is intensified many times over.

Malleable: Capable of being hammered into thin sheets.

Optical fiber: A thin strand of glass through which light passes; the light carries a message, much as an electric current carries a message through a telephone wire.

Periodic table: A chart that shows how chemical elements are related to each other.

Radioactive isotope: An isotope that breaks apart and gives off some form of radiation.

Stable: Not likely to react with other materials.

Gadolinite was full of surprises. As chemists analyzed the new mineral, they found nine new elements. No mineral had ever produced such a wealth of new information.

One of the first scientists to work on gadolinite was Swedish chemist Carl Gustav Mosander (1797–1858). Mosander was able to separate gadolinite into three parts: yttria, terbia, and erbia. Later, the parts he called erbia and terbia were given different names.

In 1843, Mosander found that erbia was an entirely new substance. It consisted of a new element combined with **oxygen**. He called the new element erbium. The name came from the town near where it had been found, Ytterby. Interestingly, three other elements were also named after this small town: **terbium, yttrium,** and **ytterbium.**

Although Mosander is given credit for discovering erbium, he saw only erbia, the compound of erbium and oxygen. The erbia he saw was not even pure, but was mixed with other rare earth element oxides.

The first pure samples of erbium oxide (erbia) were produced in 1905 by French chemist George Urbain (1872–1938) and American chemist Charles James (1880–1928). It was not until 1934 that the first pure erbium metal was produced.

Physical Properties

Erbium metal has a bright, shiny surface, much like metallic silver. It is soft and malleable. Malleable means capable of being hammered into thin sheets. Erbium has a melting point of 2,784°F (1,529°C) and a boiling point of about 5,194°F (2,868°C). Its density is 9.16 grams per cubic centimeter.

Chemical Properties

Erbium is fairly stable in air. It does not react with oxygen as quickly as most other lanthanoids. Erbium compounds tend to be pink or red. They are sometimes used to color glass and ceramics.

Occurrence in Nature

Erbium ranks about number 42 in abundance in Earth's crust. It is more common than **bromine**, **uranium**, **tin**, **silver**, and **mercury**. It occurs in many different rare earth minerals, naturally occurring lanthanoid mixtures. Some common sources of erbium are xenotime, fergusonite, gadolinite, and euxenite.

Isotopes

Six naturally occurring stable isotopes of erbium are known. Isotopes are two or more forms of an element. Isotopes differ from each other according to their mass number. The number written to the right of the element's name is the mass number. The mass number represents the number of protons plus neutrons in the nucleus of an atom of the element. The number of protons determines the element, but the number of neutrons in the atom of any one element can vary. Each variation is an isotope. The naturally occurring isotopes of erbium are erbium-162, erbium-164, erbium-166, erbium-167, erbium-168, and erbium-170.

Thirty radioactive isotopes of erbium are known also. A radioactive isotope is one that breaks apart and gives off some form of radiation. Radioactive isotopes are produced when very small particles, such as protons or neutrons, are fired at atoms. These particles stick in the atoms and make them radioactive. None of the radioactive isotopes of erbium has any important uses.

Extraction

Erbium in a mineral is first converted into erbium fluoride (ErF$_3$). Pure erbium is then obtained by passing an electric current through molten (melted) erbium fluoride:

$$2ErF_3 \rightarrow 2Er + 3F_2$$

Uses

Erbium metal has few uses. It is sometimes alloyed with **vanadium** metal. An alloy is made by melting and mixing two or more metals. The mixture has properties different than those of the individual metals. A vanadium-erbium alloy is easier to work with than pure vanadium metal.

The most important uses of erbium are in lasers and optical fibers. A laser is a device for making very intense light of one specific color. The light is intensified and focused into a narrow beam that can cut through metal. Lasers have many practical applications.

Erbium lasers are used to treat skin problems. The lasers have been used to remove wrinkles and scars. They work better than other kinds of lasers because they do not penetrate the skin very deeply. They also produce little heat and cause few side effects.

An optical fiber is a thin thread-like piece of glass or plastic through which light travels easily. Light carries messages along the fiber, much as

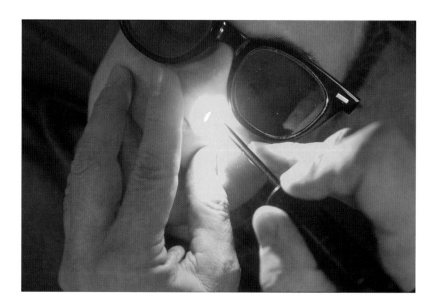

An important use of erbium is in lasers. Here, a patient undergoes cosmetic laser surgery. © WILL & DENI MCINTYRE/PHOTO RESEARCHERS, INC.

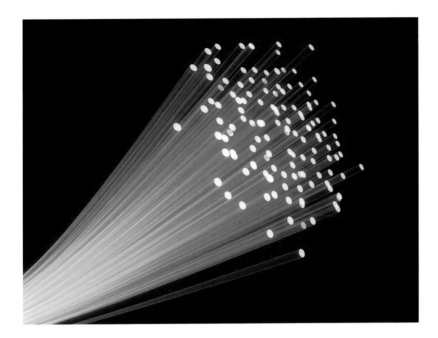

electricity carries messages along **copper** telephone wires. Erbium optical fibers carry messages in long distance communication systems and in military applications. Telephone providers have converted various copper phone lines to optical fibers for improved clarity. Optical fibers carry far more information than the old bundles of copper.

Optical fibers have proven to be an ideal method of transmitting high-definition television (HDTV) signals. HDTV has become one of the most popular new technology items for consumers. Since HDTVs were introduced in the late 1990s, sales of the sets have increased rapidly. By some estimates, the sales of HDTVs will reach 250 million by the year 2015.

The signals in HDTVs contain twice as much "information" as conventional TVs do. These signals result in a much clearer picture. However, the picture on an HDTV looks just like a regular TV unless optical fibers are used. With optical fibers, the HDTV signal can transmit a nearly perfect image.

Compounds

There are no commercially important erbium compounds.

Health Effects

Almost nothing is known about the health effects of erbium on plants, humans, or other animals. In such cases, it is usually safest to assume that the element is toxic.

Europium

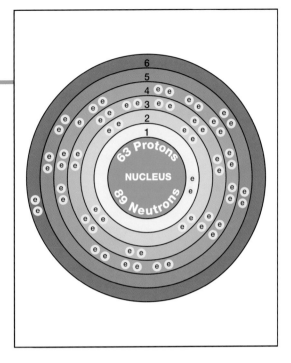

Overview

Europium was discovered in 1901 by French chemist Eugène-Anatole Demarçay (1852–1904). Demarçay named the element after the continent of Europe. It was one of the last of the rare earth elements discovered.

The rare earth elements include numbers 57 to 71 in Row 6 of the periodic table. The periodic table is a chart that shows how chemical elements are related to each other. A better name for these elements is lanthanoids. This name comes from the element **lanthanum**. Rare earth elements are not especially rare, but were originally very difficult to separate from one another.

Europium is the most active of the lanthanoids. It is more likely to react with other elements than the other rare earth elements.

Europium is quite expensive to produce, so it has few practical uses. It is used in television tubes and lasers.

Discovery and Naming

In 1901, Demarçay was studying **samarium**, a new element that had been discovered some 20 years earlier. In his studies, Demarçay made

Key Facts

Symbol: Eu

Atomic Number: 63

Atomic Mass: 151.964

Family: Lanthanoid (rare earth metal)

Pronunciation: yuh-RO-pea-um

WORDS TO KNOW

Isotopes: Two or more forms of an element that differ from each other according to their mass number.

Fission: The process in which large atoms break apart, releasing large amounts of energy, smaller atoms, and neutrons in the process.

Lanthanoids (rare earth elements): The elements in the periodic table with atomic numbers 57 through 71.

Periodic table: A chart that shows how chemical elements are related to each other.

Phosphor: A material that gives off light when struck by electrons.

Radioactive isotope: An isotope that breaks apart and gives off some form of radiation.

Toxic: Poisonous.

an interesting discovery. The new element was not one, but two elements. Demarçay gave the original name of samarium to one, and the other he called europium, after the continent of Europe.

More than a century earlier, a heavy new mineral had been found near the town of Bastnas, Sweden, and given the name cerite.

Chemists found that cerite was a complex material. One hundred years of research revealed seven new elements in cerite. Europium was the last of these new elements to be identified.

Physical Properties

Europium has a bright, shiny surface. It is steel gray and has a melting point of 1,520°F (826°C) and a boiling point of about 2,784°F (1,529°C). The density is 5.24 grams per cubic centimeter, about five times the density of water.

Europium has a strong tendency to absorb neutrons, making it useful in nuclear power production. A nuclear power plant produces electricity from the energy released by nuclear fission. Slow-moving neutrons collide with **uranium** or **plutonium** atoms, breaking them apart and releasing energy as heat. The amount of energy produced in a nuclear power plant is controlled by the number of neutrons present. Europium is used to absorb neutrons in this kind of control system.

Chemical Properties

Europium is the most active of the lanthanoids. It reacts vigorously with water to give off **hydrogen**. It also reacts with **oxygen** in the air, catching fire spontaneously. Scientists must use great care in handling the metal.

Occurrence in Nature

Europium is not abundant in Earth's surface. It is thought to occur at a concentration of no more than about one part per million. That makes it one of the least abundant of the rare earth elements. The study of light from the sun and certain stars indicates that europium is present in these bodies as well.

The most common ores of europium are monazite, bastnasite, and gadolinite.

Isotopes

Two naturally occurring stable isotopes of europium exist: europium-151 and europium-153. Isotopes are two or more forms of an element. Isotopes differ from each other according to their mass number. The number written to the right of the element's name is the mass number. The mass number represents the number of protons plus neutrons in the nucleus of an atom of the element. The number of protons determines the element, but the number of neutrons in the atom of any one element can vary. Each variation is an isotope.

Thirty-six radioactive isotopes of europium have also been made artificially. A radioactive isotope is one that breaks apart and gives off some form of radiation. Radioactive isotopes are produced when very small particles are fired at atoms. These particles stick in the atoms and make them radioactive. None of the radioactive isotopes of europium has any commercial use.

Extraction

Europium is prepared by heating its oxide with lanthanum metal:

$$Eu_2O_3 + 2La \rightarrow La_2O_3 + 2Eu$$

Europium metal is quite expensive to make.

Uses

There are no commercially important uses for europium metal.

Compounds

The most common use of europium compounds is in making phosphors. A phosphor is a material that shines when struck by electrons. The color of the phosphor depends on the elements from which it is made. Phosphors containing europium compounds give off red light. The red color on a television screen, for example, may be produced by phosphors containing europium oxide. It is also used in energy-efficient LED lightbulbs. Europium oxide is a compound made of europium metal and oxygen. In 2007 europium oxide sold for about $1,200 per kilogram.

Europium oxide phosphors are also used in printing postage stamps. These phosphors make it possible for machines to "read" a stamp and know what its value is. If the wrong stamp is on a letter, the machine can tell from reading the phosphor. The machine will then send the letter back to the person who mailed it.

Health Effects

Except for europium's tendency to catch fire, little information is available on its health effects. In general, it is regarded as toxic and must be handled with great caution.

Fermium

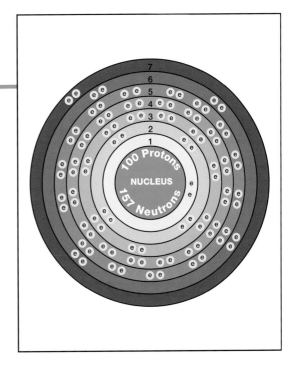

Overview

Fermium is one of the transuranium elements, which lie beyond **uranium** in the periodic table. The periodic table is a chart that shows how chemical elements are related to each other. Uranium is element number 92, so all elements with larger atomic numbers are transuranium elements.

Discovery and Naming

Fermium was discovered in 1952, among the products formed during the first hydrogen bomb test at Eniwetok Atoll, Marshall Islands, in the Pacific Ocean. For security reasons, this discovery was not announced until 1955. Credit for the discovery of fermium goes to a group of University of California scientists under the direction of Albert Ghiorso (1915–). The element was named for Italian physicist Enrico Fermi (1901–1954). Fermi, who made many important scientific discoveries in his life, was a leader of the U.S. effort to build the world's first fission (atomic) bomb during World War II.

Key Facts

Symbol: Fm

Atomic Number: 100

Atomic Mass: [257]

Family: Actinoid; transuranium element

Pronunciation: FER-me-um

WORDS TO KNOW

Actinoid family: Elements with atomic numbers 89 through 103.

Half life: The time it takes for half of a sample of a radioactive element to break down.

Isotopes: Two or more forms of an element that differ from each other according to their mass number.

Nuclear reactor: A device in which nuclear reactions occur.

Periodic table: A chart that shows how the chemical elements are related to each other.

Radioactive: Having a tendency to give off radiation.

Transuranium element: An element with an atomic number greater than 92.

Physical and Chemical Properties

Too little fermium has been prepared to allow scientists to determine most of its physical and chemical properties. Its melting point has been determined to be 2,780°F (1,527°C).

Occurrence in Nature

Fermium does not occur naturally in Earth's crust.

Isotopes

All 20 known isotopes of fermium are radioactive. The most stable isotope is fermium-257. Isotopes are two or more forms of an element. Isotopes differ from each other according to their mass number. The number written to the right of the element's name is the mass number. The mass number represents the number of protons plus neutrons in the nucleus of an atom of the element. The number of protons determines the element, but the number of neutrons in the atom of any one element can vary. Each variation is an isotope.

The half life of fermium-257 is 100.5 days. The half life of a radioactive element is the time it takes for half of a sample of the element to break down. A radioactive isotope is one that breaks apart and gives off some form of radiation. For example, suppose 100 grams of fermium-257 is made. Fifty grams of the isotope would be left about 100 days later. After another 100 days, only 25 grams of the isotope would remain.

Facts about Enrico Fermi

Fermium's namesake Enrico Fermi taught physics and did research at the University of Rome from 1926 to 1938. During this period, he learned how to use neutrons to change elements from one form (isotope) to another. He received the Nobel Prize in physics in 1938 for these discoveries.

The year he received the Nobel Prize was a difficult time in Europe. Benito Mussolini (1883–1945) had just come to power in Italy. Mussolini followed many of the same unjust policies as Nazi leader Adolf Hitler (1889–1945) did in Germany. One of these policies was anti-Semitism (hostility toward Jews). Fermi, whose wife was Jewish, began to worry about what might happen if they stayed in Italy. Like many other scientists in Europe, he decided to come to the United States, where he took a job at Columbia University in New York.

Soon after Fermi arrived in the United States, he discovered that his experience with neutrons was very valuable. The U.S. government had undertaken a huge top-secret research program called the Manhattan Project. The purpose of the Manhattan Project was to find a way to build an atomic bomb. Fermi was placed in charge of one part of that project.

Fermi's responsibility was to study the reaction that occurs when uranium atoms are bombarded by neutrons. His team did most of this research at the University of Chicago. On December 2, 1942, the team made an important breakthrough. They produced the first self-sustaining chain reaction in history. A self-sustaining chain reaction is one in which neutrons split uranium atoms apart. Large amounts of energy are produced in the reaction. Additional neutrons are also formed. These neutrons can be used to make the reaction repeat over and over again. The reaction eventually formed the basis of the first atomic bombs built three years later.

After the war, Fermi returned to the University of Chicago as professor of physics. He died at the early age of 53 of stomach cancer. In his honor, the U.S. government created the Enrico Fermi Award for accomplishments in nuclear physics.

Extraction

Fermium is not extracted from Earth's crust.

Uses

Fermium is sometimes used in scientific research, but it has no commercial applications.

Compounds

There are no commercially important compounds of fermium.

Enrico Fermi.

Health Effects

Scientists know too little about fermium to be aware of its health effects. As a radioactive element, however, it does pose a threat to human health.

Fluorine

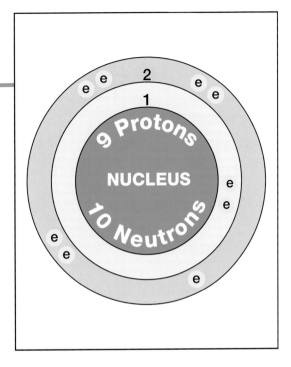

Overview

Fluorine is the lightest member of the halogen family, elements in Group 17 (VIIA) of the periodic table. The periodic table is a chart that shows how elements are related to one another. The halogens also include **chlorine**, **bromine**, **iodine**, and **astatine**. Fluorine is the most active chemical element, reacting with virtually every element. It even reacts with the noble gases at high temperatures and pressures. The noble gases, Group 18 (VIIIA) in the periodic table, are also known as the inert gases. They generally do not react with other elements.

Fluorine was discovered in 1886 by French chemist Henri Moissan (1852–1907). Moissan collected the gas by passing an electric current through one of its compounds, hydrogen fluoride (H_2F_2).

One of the best known uses of fluorine is in the production of fluorides, used as additives in toothpastes and municipal water supplies. Fluorides are effective in preventing tooth decay and have been widely used in the United States for this purpose since the 1950s.

Another well known group of fluorine compounds is the chlorofluorocarbons (CFCs). For many years, the CFCs were used for a wide variety of industrial purposes, including refrigeration, cleaning systems, and as

Key Facts

Symbol: F

Atomic Number: 9

Atomic Mass: 18.9984032

Family: Group 17 (VIIA); halogen

Pronunciation: FLOR-een

WORDS TO KNOW

Halogen: One of the elements in Group 17 (VIIA) of the periodic table.

Isotopes: Two or more forms of an element that differ from each other according to their mass number.

Noble gas: An element in Group 18 (VIIIA) of the periodic table.

Ozone: A form of oxygen that filters out harmful radiation from the sun.

Ozone layer: The layer of ozone that shields Earth from harmful ultraviolet radiation from the sun.

Periodic table: A chart that shows how chemical elements are related to each other.

Polymerization: The process by which many thousands of individual tetrafluoroethlylene (TFE) molecules join together to make one very large molecule.

Radioactive isotope: An isotope that breaks apart and gives off some form of radiation.

Serendipity: Discovering something of value when not seeking it; for example, making a discovery by chance or accident.

Toxic: Poisonous.

Ultraviolet (UV) radiation: Electromagnetic radiation (energy) of a wavelength just shorter than the violet (shortest wavelength) end of the visible light spectrum and thus with higher energy than visible light.

propellants in aerosol products. They were considered to be one of the most successful family of synthetic compounds ever invented.

However, CFCs react with ozone (O_3) in the upper atmosphere. Scientists discovered that the CFCs were depleting the ozone layer. The ozone layer filters harmful ultraviolet (UV) radiation from the sun. Ultraviolet radiation is electromagnetic radiation (energy) of a wavelength just shorter than the violet (shortest wavelength) end of the visible light spectrum and thus with higher energy than visible light. To protect the ozone layer, the production of CFCs was banned in the United States and many other countries.

Discovery and Naming

Chemistry has always been a dangerous science. Early chemistry was a hazardous occupation. Men and women worked with chemicals about which they knew very little. The discovery of new compounds and elements could easily have tragic consequences.

Fluorine was particularly vicious. Chemists suffered terrible injuries and even died in their attempts to study the element. Fluorine gas is extremely damaging to the soft tissues of the respiratory tract.

In the early 1500s, German scholar Georgius Agricola (1494–1555) described a mineral he called fluorspar. The name fluorspar comes from the Latin word *fluere,* meaning "to flow." Agricola claimed that fluorspar added to molten metallic ores made them more liquid and easier to work with. Although Agricola did not realize it, fluorspar is a mineral of fluorine and contains calcium fluoride (CaF_2).

Fluorspar became the subject of intense study by early chemists. In 1670, German glass cutter Heinrich Schwanhard discovered that a mixture of fluorspar and acid formed a substance that could be used to etch glass. Etching is a process by which a pattern is drawn into glass. The chemical reaction leaves a frosted image. Etching is used to produce artistic shapes on glass as well as in the manufacture of precise scientific measuring instruments.

The new etching material was identified in 1771 by Swedish chemist Carl Wilhelm Scheele (1742–1786). Scheele described, in detail, the properties of this material, hydrofluoric acid (H_2F_2). His work set off an intense study of the acid and its composition.

One goal was to find ways to break hydrofluoric acid into elements. Chemists suspected that one element had never been seen before. Little did they know, however, what a dangerous new element it would be. During studies of hydrofluoric acid, many chemists were disabled when they inhaled hydrogen fluoride gas. One chemist, Belgian Paulin Louyet (1818–1850), died from his exposure to the chemical.

Finally, in 1888, the problem was solved. Moissan made a solution of hydrofluoric acid in potassium hydrogen fluoride (KHF_2). He then cooled the solution to –9.4°F (–23°C) and passed an electric current through it. A gas appeared at one end of the apparatus. He gave the name fluorine to the new element. The name comes from the mineral fluorspar.

Physical Properties

Fluorine is a pale yellow gas with a density of 1.695 grams per liter. That makes fluorine about 1.3 times as dense as air. Fluorine changes from a gas to a liquid at a temperature of –306.5°F (–188.13°C) and from a liquid to a solid at –363.30°F (–219.61°C).

Fluorine has a strong and characteristic odor that can be detected in very small amounts, as low as 20 parts per billion. This property is very helpful to those who work with fluorine. It means that the gas can be detected and avoided in case it leaks into a room.

Chemical Properties

Fluorine is the most reactive nonmetallic element. It combines easily with almost every other element. Compounds of fluorine and the noble gases have even been made. The noble gases are the elements in Group 18 (VIIIA) of the periodic table. They are normally very unreactive. Fluorine also reacts with most compounds, often violently. For example, when mixed with water, it reacts explosively. For these reasons, it must be handled with extreme care in the laboratory.

Occurrence in Nature

Fluorine never occurs as a free element in nature. The most common fluorine minerals are fluorspar, fluorapatite, and cryolite. Apatite is a complex mineral containing primarily **calcium**, **phosphorus**, and **oxygen**, usually with fluorine. Cryolite is also known as Greenland spar. (The country of Greenland is the only commercial source of this mineral.) It consists primarily of sodium **aluminum** fluoride (Na_3ALF_6). The major sources of fluorspar are China, Mexico, Mongolia, and South Africa. In 2008 in the United States, fluorspar was produced as a by-product of limestone quarrying in Illinois. The United States imports most of the fluorspar it needs from China and Mexico.

Fluorine is an abundant element in Earth's crust, estimated at about 0.06 percent in the earth. That makes it about the 13th most common element in the crust. It is about as abundant as **manganese** or **barium**.

Isotopes

Only one naturally occurring isotope of fluorine (fluroine-19) is known to exist. Isotopes are two or more forms of an element. Isotopes differ from each other according to their mass number. The number written to the right of the element's name is the mass number. The mass number represents the number of protons plus neutrons in the nucleus of an atom of the element. The number of protons determines the element, but the number of neutrons in the atom of any one element can vary. Each variation is an isotope.

In addition, 10 artificial radioactive isotopes of fluorine are known. A radioactive isotope is one that breaks apart and gives off some form of radiation. Radioactive isotopes are produced when very small particles are fired at atoms. These particles stick in the atoms and make them radioactive.

One radioactive isotope, fluorine-18, is sometimes used for medical studies. It is combined chemically with glucose (blood sugar) molecules and injected into the body. It then travels to cancer cells, which reproduce rapidly and use glucose very quickly. The presence of fluorine-18 can be detected in these cells because of the radiation it gives off. Using this method, medical workers can determine the location, size, and other properties of cancerous cells in the body.

Extraction

Fluorine is made commercially by Moissan's method. An electric current is passed through a mixture of hydrogen fluoride and potassium hydrogen fluoride:

$$H_2F_2 \xrightarrow{\text{electric current}} H_2 + F_2$$

Uses and Compounds

Fluorine has relatively few uses as an element. It is much too active for such applications. One use of elemental fluorine is in rocket fuels. It helps other materials burn, like oxygen does. The greatest majority of fluorine is used to make compounds of fluorine.

Fluorides are compounds of fluorine with (usually) one other element. **Sodium** fluoride (NaF), calcium fluoride (CaF_2), and stannous fluoride (SnF_2) are examples of fluorides.

A familiar use of some fluoride compounds is in toothpastes. Studies show that small amounts of fluorides can help reduce tooth decay. Fluorides are deposited as new tooth material is formed, making it strong and resistant to decay.

Some cities add fluorides to their water supply. By doing so, they hope to improve the dental health of everyone living in the city. Young people, whose teeth are still developing, benefit the most. The process of adding fluorides to public water supplies is called fluoridation. Too much fluorine in the water leads to a light brown and permanent staining of teeth.

Some people worry about the long-term health effect of fluorides added to public water supplies. They point out that fluorine is a deadly poison and that fluorides can be toxic as well. It is true that fluorine gas is very toxic, but the properties of compounds are different from the elements involved. Little evidence exists to support these concerns.

Fluorides tend to be dangerous only in large doses. The amount of fluoride added to public water supplies is usually very small, only a few parts per million. Most dental and health experts believe that fluoridation is a helpful public health practice, not a threat to the health of individuals.

Chlorofluorocarbons (CFCs) At one time, another major use of fluorine was in the production of CFCs. CFCs were discovered in the late 1920s by American chemical engineer Thomas Midgley Jr. (1889–1944). These compounds have a number of interesting properties. They are very stable and do not break down when used in a variety of industrial operations. They were widely used in cooling and refrigeration systems, as cleaning agents, in aerosol sprays, and in specialized polymers. The production of CFCs grew from about 1 million kilograms in 1935 to more than 300 million kilograms in 1965 to more than 700 million kilograms in 1985.

By the mid-1980s, however, evidence showed that the compounds were damaging Earth's ozone layer. This layer lies at an altitude of 12 to 30 miles (20 to 50 kilometers) above Earth's surface. It is important to life because it shields Earth from the sun's harmful ultraviolet radiation. As a result, CFCs were phased out. The compounds are no longer produced

Accidents Happen!

Serendipity plays a big part in scientific research. The term serendipity means a discovery made by accident. One of the most profitable discoveries made this way is the material Teflon. The name Teflon is the trade name of a type of plastic made by the DuPont Chemical Company. It has become an important commercial product for one main reason: very few things stick to Teflon. Most kitchen cupboards probably contain skillets and other pans with cooking surfaces covered with Teflon. Most food will not stick to Teflon-covered pans as it cooks. And foods cooked in Teflon pans need no oil or butter.

Teflon was discovered by accident in 1938 by a DuPont chemist named Roy Plunkett (1911–1994). Plunkett was working on the development of chlorofluorocarbons (CFCs) for DuPont. He wanted to see what happened when one compound, tetrafluoroethylene, or TFE (C_2F_4), was mixed with hydrochloric acid (HCl). To carry out the experiment, he set up the equipment so that the gaseous TFE would flow into a container of HCl.

When he opened the valve on the TFE container, however, nothing came out. Plunkett could have discarded the tank, but he did not. Instead, he sawed open the container. Inside he found that the TFE had polymerized into a single mass. Polymerization is the process by which many thousands of individual TFE molecules join together to make one very large molecule. The large molecule is called polytetrafluorethylene, or PTFE.

Plunkett scraped the white PTFE powder out and sent it to DuPont scientists working on artificial fibers. The scientists studied the properties of PTFE. They discovered its non-stick qualities and were soon working on a number of applications for the new material.

DuPont registered the Teflon trademark in 1945 and released its first Teflon products a year later. Since then the non-stick coating has become a household staple on kitchen cookware, in baking sprays, and as a stain repellant for fabrics and textiles.

or used in the United States and most other parts of the world. Safer substitutes are now being used in the products that once used CFCs.

People are often confused about CFCs. They were popular industrial chemicals because they do not break down very easily. They were long used in most car air conditioners as a heat transfer fluid. CFCs took heat out of the car and moved it into the outside air. This process was carried out over and over again. CFCs in a car's system lasted a very long time.

But eventually scientists realized the threat that CFCs posed to the ozone layer *because* they break down. How can that be?

There is always a little leakage from auto air conditioning units and from every device in which CFCs are used. CFCs are gases or liquids that easily vaporize and float upward into the atmosphere, eventually reaching the ozone layer. At this height above Earth's surface, CFCs encounter intense radiation from the sun and break apart. A molecule that was stable near Earth's surface has now become *un*stable.

When a CFC molecule breaks apart, it forms a single chlorine atom:

$$CFC \xrightarrow{\text{sunlight}} CFC^* + Cl$$

(The CFC* indicates a CFC molecule without one chlorine atom.) The chlorine formed in this reaction can react with an ozone (O_3) molecule:

$$Cl + O_3 \rightarrow ClO + O_2$$

Here's the problem: The ozone (O_3) will filter out harmful radiation from the sun. It removes most of the radiation that causes severe sunburns and skin cancer. But oxygen (O_2) cannot do the same thing. Therefore, (1) the more CFCs in the atmosphere, the more chlorine atoms; (2) the more chlorine atoms, the fewer ozone molecules; (3) the fewer ozone molecules, the more harmful radiation reaching Earth's surface; and (4) the more harmful radiation, the more cases of skin cancer and other health problems.

It is this series of events that convinced most nations to ban the further production and use of CFCs.

Health Effects

As noted under "Discovery and Naming," fluorine can be quite dangerous. If inhaled in small amounts, it causes severe irritation to the respiratory system (nose, throat, and lungs). In larger amounts, it can cause death. The highest recommended exposure to fluorine is one part per million of air over an eight-hour period.

Francium

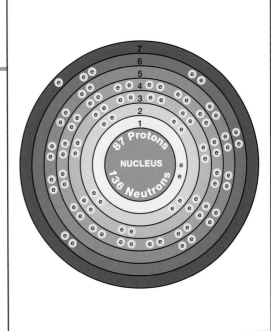

Overview

Francium is an alkali metal, a member of Group 1 (IA) in the periodic table. The periodic table is a chart that shows how chemical elements are related to each other.

Francium may be the rarest element found on Earth's surface. Some experts believe that no more than 15 grams (less than an ounce) of the element exist in Earth's crust. The element was discovered in 1939 by French chemist Marguerite Perey (1909–1975). All isotopes of francium are radioactive.

Key Facts

Symbol: Fr

Atomic Number: 87

Atomic Mass: [223]

Family: Group 1 (IA); alkali metal

Pronunciation: FRAN-see-um

Discovery and Naming

Francium was one of the last naturally occurring elements to be discovered. Chemists had been searching for it since the development of the periodic table.

In the early 1900s, nearly all boxes on the periodic table had been filled. One element had been found to fit into each box. By the 1930s, only three remained empty—elements with atomic numbers of 43, 85, and 87.

WORDS TO KNOW

Isotopes: Two or more forms of an element that differ from each other according to their mass number.

Particle accelerator ("atom smasher"): A machine used to get small particles, like protons, moving very rapidly.

Periodic table: A chart that shows how chemical elements are related to each other.

Radiation: Energy transmitted in the form of electromagnetic waves or subatomic particles.

Radioactive isotope: An isotope that breaks apart and gives off some form of radiation.

The search for the three remaining elements produced a number of incorrect results. For example, American chemist Fred Allison (1882–1974) announced the discovery of elements 85 and 87 in 1931. He suggested the names of alabamine and virginium, in honor of the states in which he was born (Virginia) and worked (Alabama). But other scientists were not able to confirm Allison's discoveries.

Element 87 was isolated by Marguerite Perey, who was studying the radioactive decay of the element **actinium**. Radioactive elements like actinium break apart spontaneously, giving off energy and particles. This process results in the formation of new elements.

Perey found that 99 percent of all actinium atoms decay into **thorium**. The remaining 1 percent breaks down into a new element, number 87.

Perey's Legacy Marguerite Perey made important discoveries during an era when few women held prominent roles in the sciences. She was interested in science even as a small child. However, her father died early on, and there was no money for Perey to attend a university. Instead, she found a job at the Radium Institute in Paris. The Radium Institute had been founded by Marie Curie (1867–1934) and her husband, Pierre Curie (1859–1906), to study radioactive materials.

Perey was originally hired for a three-month period. But Madame Curie was very impressed with Perey's skills in the laboratory. Perey eventually ended up working at the Radium Institute until 1935.

One of the projects Perey worked on was the radioactive decay of actinium. When actinium decays, it gives off radiation and changes into another element, thorium. Thorium, in turn, also gives off radiation and changes into another element, **radium**. This process is repeated a number of times. In each step, a radioactive element decays to form another element.

As Perey studied this series of reactions, she made an interesting discovery. The mixture of elements that are formed in these reactions contained a substance she did not recognize. She decided to find out what that substance was. She was eventually able to show that it was a new element, with atomic number 87. The element was one of the last naturally occurring elements to be discovered. Perey named the element in honor of her native land, France.

Perey was the first woman ever elected to the French Academy of Science. Even Marie Curie had not earned that honor. Perey died in 1975 after a 15-year battle with cancer.

Physical and Chemical Properties

Until very recently, there was not enough francium to permit a study of its properties. In 1991, scientists at the State University of New York at Stony Brook developed a method for making small amounts of francium and holding them

French physicist Marguerite Perey. AIP/PHOTO RESEARCHERS, INC.

in a "laser trap" for up to 30 seconds. A laser trap is made by focusing a number of laser beams at a bunch of atoms, holding them motionless so that they can be studied. This research confirmed the fact that francium is similar to the other alkali metals above it on the periodic table. The alkali metals are the elements in Group 1. Francium's melting point has been determined to be 81°F (27°C), and its boiling point is estimated to be about 1,250°F (677°C).

Occurrence in Nature

Francium is now produced in particle accelerators, which are also called atom smashers. These machines accelerate small particles, like protons, to nearly the speed of light, 186,000 miles per second (300,000 kilometers per second). The particles collide with target atoms, such as **copper**, **gold**, or **tin**. Target atoms fragment, forming new elements and particles.

Isotopes

Forty-one isotopes of francium have been produced artificially. The most stable is francium-223. Isotopes are two or more forms of an element. Isotopes differ from each other according to their mass number. The number written to the right of the element's name is the mass number. The mass number represents the number of protons plus neutrons in the nucleus of an atom of the element. The number of protons determines the element, but the number of neutrons in the atom of any one element can vary. Each variation is an isotope.

Francium-223 has a half life of 21.8 minutes. The half life of a radioactive element is the time it takes for half of a sample of the element to break down. That means that 100 grams of francium-223 will break down so that only 50 grams are left after 21.8 minutes. Another 21.8 minutes later, 25 grams of francium-223 will remain, and so on.

Extraction

Francium is not extracted from Earth's crust.

Uses

Francium has no uses because of its rarity. Scientists hope to learn about the composition of matter by studying the element, however.

Compounds

There are no commercially important compounds of francium.

Health Effects

Scientists know too little about francium to be aware of its health effects. As a radioactive element, however, it does pose a threat to human health.

Where to Learn More

The following list focuses on works written for readers of middle school and high school age. Books aimed at adult readers have been included when they are especially important in providing information or analysis that would otherwise be unavailable, or because they have become classics.

Books

Atkins, P. W. *The Periodic Kingdom: A Journey into the Land of the Chemical Elements.* New York: HarperCollins, 1997.

Ball, Philip. *The Ingredients: A Guided Tour of the Elements.* New York: Oxford University Press, 2003.

Emsley, John. *Nature's Building Blocks: An A-Z Guide to the Elements.* New York: Oxford University Press, 2003.

Gray, Theodore W., and Nick Mann. *The Elements: A Visual Exploration of Every Known Atom in the Universe.* New York: Black Dog & Leventhal Publishers, 2009.

Greenwood, N. N., and A. Earnshaw. *Chemistry of the Elements, 2nd Edition.* Oxford: Pergamon Press, 2006.

Harvey, David I. *Deadly Sunshine: The History and Fatal Legacy of Radium.* Stroud, Gloucestershire, UK: Tempus, 2005.

Krebs, Robert E. *The History and Use of Our Earth's Chemical Elements: A Reference Guide, 2nd Edition.* Westport, CT: Greenwood Publishing Group, 2006.

Lew, Kristi. *Radium.* New York: Rosen Central, 2009.

Lewis, Richard J., Sr. *Hawley's Condensed Chemical Dictionary, 15th edition.* New York: Wiley Interscience, 2007.

Miller, Ron. *The Elements.* Minneapolis: Twentieth Century Books, 2006.

Minerals Information Team. *Minerals Yearbook.* Reston, VA: U.S. Geological Survey, 2009.

Nachaef, N. *The Chemical Elements: The Exciting Story of Their Discovery and of the Great Scientists Who Found Them.* Jersey City, NJ: Parkwest Publications, 1997.

Ruben, Samuel. *Handbook of the Elements, 3rd Edition.* La Salle, IL: Open Court Publishing Company, 1999.

Stwertka, Albert. *A Guide to the Elements, 2nd Edition.* New York: Oxford University Press, 2002.

Weeks, Mary Elvira. *Discovery of the Elements.* Whitefish, MT: Kessinger Publishing Reprints, 2003.

Periodicals

Barber, R. C., et al. "Discovery of the Element with Atomic Number 112." *Pure and Applied Chemistry* (August 13, 2009): 1,331–1,343.

Barceloux, D. G. "Chromium." *Journal of Toxicology. Clinical Toxicology* 37 (1999): 173–194.

Biever, Celeste. "The Tick, Tock of a Strontium Clock Gets Steadier." *New Scientist* (December 9, 2006): 32.

Camparo, James. "The Rubidium Atomic Clock and Basic Research." *Physics Today* (November 2007): 33–39.

Cotton, S. A. "The Actinides." *Annual Reports Section A (Inorganic Chemistry)* 105 (2009): 287–296.

Cotton, S. A. "Titanium, Zirconium, and Hafnium." *Annual Reports Section A (Inorganic Chemistry)* 105 (2009): 177–187.

Crans, Debbie D., Jason J. Smee, Ernestas Gaidamauskas, and Luqin Yang. "The Chemistry and Biochemistry of Vanadium and the Biological Activities Exerted by Vanadium Compounds." *Chemical Reviews* (January 2004): 849–902.

Croswell, Ken. "Fluorine: An Element-ary Mystery." *Sky and Telescope* (September 2003): 30–37.

Fricker, Simon P. "Therapeutic Applications of Lanthanides." *Chemical Society Reviews* 35 (2006): 524–533.

Garelick, Hemda, Huw Jones, Agnieszka Dybowska, and Eugenia Valsami-Jones. "Arsenic Pollution Sources." *Reviews of Environmental Contamination and Toxicology* 197 (2008): 17–60.

Geim, Andre K., and Philip Kim. "Carbon Wonderland." *Scientific American* (April 2008): 90–97.

Gomez, Gerda E. "Lithium Treatment: Present and Future." *Journal of Psychosocial Nursing and Mental Health Services* (August 2001): 31–37.

Gruber, Nicolas, and James N. Galloway. "An Earth-system Perspective of the Global Nitrogen Cycle." *Nature* (January 17, 2008): 293–296.

Guerrera, Mary P., Stella Lucia Volpe, and James Mao Jun. "Therapeutic Uses of Magnesium." *American Family Physician* (July 15, 2009): 157–162.

Kolanz, Marc. "Beryllium History and Public Policy." *Public Health Reports* (July 2008): 423–428.

Ljung, Karin, and Marie Vahter. "Time to Re-Evaluate the Guideline Value for Manganese in Drinking Water?" *Environmental Health Perspectives* (November 2007): 1,533–1,538.

Martin, Richard. "Uranium Is So Last Century—Enter Thorium, the New Green Nuke." *WIRED* (January 2010).

Nielsen, Forrest H. "Is Boron Nutritionally Relevant?" *Nutrition Reviews* (April 2008): 183–191.

Ogden, Joan. "High Hopes for Hydrogen." *Scientific American* (September 2006): 94–101.

Parr, J. "Carbon, Silicon, Germanium, Tin and Lead." *Annual Reports Section A (Inorganic Chemistry)* 105 (2009): 117–139.

Roney, Nicolette, et al. "ATSDR Evaluation of Potential for Human Exposure to Zinc." *Toxicology and Industrial Health* 22 (2006): 423–493.

Satyapal, Sunita, John Petrovic, and George Thomas. "Gassing Up with Hydrogen." *Scientific American* (April 2007): 80–87.

Stein, Andreas. "Materials Science: Germanium Takes Holey Orders." *Nature* (June 29, 2006): 1,055–1,056.

Utiger, Robert D. "Iodine Nutrition—More Is Better." *New England Journal of Medicine* (June 29, 2006): 2,819–2,821.

Vaccari, David A. "Phosphorus: A Looming Crisis." *Scientific American* (June 2009): 54–59.

van Delft, Dirk. "Little Cup of Helium, Big Science." *Physics Today* (March 2008): 36–42.

Web Sites

Air Chek, Inc. "The Radon Information Center." http://www.radon.com/ (accessed February 9, 2010).

American Chemistry Council, Chlorine Chemistry Division. http://www.americanchemistry.com/chlorine/ (accessed February 9, 2010).

American Elements. "Sodium." http://www.americanelements.com/na.html (accessed February 9, 2010).

American Elements. "Zinc." http://www.americanelements.com/zn.html (accessed February 9, 2010).

BBC News. "New Element Named 'Copernicium.'" http://news.bbc.co.uk/2/hi/science/nature/8153596.stm (accessed February 9, 2010).

Borland, John. "Scientists Use Superconducting Cyclotron to Make Super-Heavy Metals" (October 24, 2007). http://www.wired.com/science/discoveries/news/2007/10/heavymetalisotopes (accessed February 9, 2010).

Canadian Nuclear Association. "How Is Nuclear Technology Used in Smoke Detectors?" http://www.cna.ca/english/pdf/NuclearFacts/18-NuclearFacts-smokedetectors.pdf (accessed February 9, 2010).

EnvironmentalChemistry.com. "Periodic Table of Elements." http://environmentalchemistry.com/yogi/periodic/ (accessed February 9, 2010).

Excalibur Mineral Company. "Tin Mineral Data." http://webmineral.com/data/Tin.shtml (accessed February 9, 2010).

GreenFacts. "Scientific Facts on Fluoride." http://www.greenfacts.org/en/fluoride/index.htm (accessed February 9, 2010).

Holden, Norman E. "History of the Origin of the Chemical Elements and Their Discoverers." http://www.nndc.bnl.gov/content/elements.html#boron (accessed February 9, 2010).

Indium Metal Information Center. http://www.indium.com/products/indiummetal.php (accessed February 9, 2010).

International Atomic Energy Agency. "Number of Reactors in Operation Worldwide." http://www.iaea.org/cgi-bin/db.page.pl/pris.oprconst.htm (accessed February 9, 2010).

International Platinum Group Metals Association. "Iridium." http://www.ipa-news.com/pgm/iridium/index.htm (accessed February 9, 2010).

International Union of Pure and Applied Chemistry (IUPAC). "The Chemical Elements." http://old.iupac.org/general/FAQs/elements.html (accessed February 9, 2010).

Lenntech. "Periodic Table." http://www.lenntech.com/periodic/periodic-chart.htm (accessed February 9, 2010).

Los Alamos National Laboratory. Chemical Division. "Periodic Table of the Elements." http://periodic.lanl.gov/default.htm (accessed February 9, 2010).

Mineral Information Institute. "Titanium." http://www.mii.org/Minerals/phototitan.html (accessed February 9, 2010).

Mineral Information Institute. "Vanadium." http://www.mii.org/Minerals/photovan.html (accessed February 9, 2010).

Nuclear Regulatory Commission. "Map of the United States Showing Locations of Operating Nuclear Power Reactors." http://www.nrc.gov/info-finder/reactor/ (accessed February 9, 2010).

ResponsibleGold.org. http://www.responsiblegold.org/ (accessed February 9, 2010).

Royal Society of Chemistry. "Radium—Ra." http://www.rsc.org/chemsoc/visualelements/Pages/data/radium_data.html (accessed February 9, 2010).

Royal Society of Chemistry. "Visual Interpretation of the Table of Elements." http://www.rsc.org/chemsoc/visualelements/index.htm (accessed February 9, 2010).

Silver Institute. "Silver Facts: Silver as an Element." http://www.silverinstitute.org/silver_element.php (accessed February 9, 2010).

Smart Elements.com. "Elements." http://www.smart-elements.com/
?arg=pse&lid=11&PHPSESSID=3760317e40a3f4cf31f2a87bc17cc413
(accessed February 9, 2010).

U.S. Department of Energy, Lawrence Berkeley National Laboratory. "Super-
heavy Element 114 Confirmed: A Stepping Stone to the Island of Stabi-
lity." http://newscenter.lbl.gov/press-releases/2009/09/24/114-confirmed/
(February 9, 2010).

U.S. Department of Health and Human Services. "Dietary Guidelines for
Americans." http://www.health.gov/dietaryguidelines/ (accessed February
9, 2010).

U.S. Environmental Protection Agency. "Carbon Dioxide." http://epa.gov/
climatechange/emissions/co2.html (accessed February 9, 2010).

U.S. Environmental Protection Agency. "A Citizen's Guide to Radon: The
Guide to Protecting Yourself and Your Family from Radon" (January
2009). http://www.epa.gov/radon/pdfs/citizensguide.pdf (accessed
February 9, 2010).

U.S. Environmental Protection Agency. "An Introduction to Indoor Air
Quality: Carbon Monoxide." http://www.epa.gov/iaq/co.html (accessed
February 9, 2010).

U.S. Geological Survey Minerals Yearbook. "Bismuth Statistics and Information." http://minerals.usgs.gov/minerals/pubs/commodity/bismuth/
(accessed February 9, 2010).

U.S. Geological Survey Minerals Yearbook. "Lithium Statistics and Information." http://minerals.usgs.gov/minerals/pubs/commodity/lithium/
(accessed February 9, 2010).

U.S. Geological Survey Minerals Yearbook. "Nickel Statistics and Information."
http://minerals.usgs.gov/minerals/pubs/commodity/nickel/ (accessed
February 9, 2010).

U.S. Geological Survey Minerals Yearbook. "Nitrogen Statistics and Information." http://minerals.usgs.gov/minerals/pubs/commodity/nitrogen/
(accessed February 9, 2010).

U.S. Geological Survey Minerals Yearbook. "Platinum-Metals Statistics and
Information." http://minerals.usgs.gov/minerals/pubs/commodity/
platinum/ (accessed February 9, 2010).

U.S. Geological Survey Minerals Yearbook. "Tin Statistics and Information."
http://minerals.usgs.gov/minerals/pubs/commodity/tin/ (accessed February
9, 2010).

U.S. Geological Survey Minerals Yearbook. "Vanadium Statistics and
Information." http://minerals.usgs.gov/minerals/pubs/commodity/vanadium/
(accessed February 9, 2010).

University of Nottingham. "The Periodic Table of Videos." http://www.
periodicvideos.com (accessed February 9, 2010).

WebQC.com. "Periodic Table of Chemical Elements." http://www.webqc.org/periodictable.php (accessed February 9, 2010).

World Information Service on Energy. "Uranium Project." http://www.wise-uranium.org/ (accessed February 9, 2010).

Organizations

Aluminum Association, 1525 Wilson Blvd., Suite 600, Arlington, VA 22209; Phone: 703-358-2960. Internet: www.aluminum.org.

American Chemistry Council, 1300 Wilson Blvd., Arlington, VA 22209. Phone: 703-741-5000. Internet: http://www.americanchemistry.com/.

American Hydrogen Association, 2350 W. Shangri La, Phoenix, AZ 85029. Phone: 602-328-4238. Internet: http://www.clean-air.org/.

American Iron and Steel Institute, 1140 Connecticut Ave., NW, Suite 705, Washington, DC 20036. Phone: 202-452-7100. Internet: http://www.steel.org.

Australian Uranium Association, GPO Box 1649, Melbourne, VIC 3001, Australia. Phone: +61 3 8616 0440. Internet: http://aua.org.au/.

Copper Development Association, 260 Madison Ave., New York, NY 10016. Phone: 212-251-7200. Internet: http://www.copper.org/.

International Atomic Energy Agency (IAEA), Vienna International Centre, PO Box 100, 1400 Vienna, Austria. Phone: (+431) 2600-0. Internet: http://www.iaea.org/.

International Lead Association. 17a Welbeck Way, London, United Kingdom, W1G 9YJ. Phone: +44 (0)207 499 8422. Internet: http://www.ila-lead.org/.

International Magnesium Association, 1000 N. Rand Rd., Suite 214, Wauconda, IL 60084. Phone: 847-526-2010. Internet: http://www.intlmag.org/.

International Molybdenum Association, 4 Heathfield Terrace, London, United Kingdom, W4 4JE. Phone: +44 (0)207 871 1580. Internet: http://www.imoa.info/.

International Platinum Group Metals Association, Schiess-Staett-Strasse 30, 80339 Munich, Germany. Phone: (+49) (0)89 519967-70. Internet: http://www.ipa-news.com/.

International Union of Pure and Applied Chemistry, IUPAC Secretariat, PO Box 13757, Research Triangle Park, NC 27709-3757. Internet: www.iupac.org/.

International Zinc Association—America, 2025 M Street, NW, Suite 800, Washington DC 20036. Phone: 202-367-1151. Internet: http://www.zinc.org/.

Los Alamos National Laboratory, PO Box 1663, Los Alamos, NM 87545. Phone: 505-667-5061. Internet: www.lanl.gov/.

Mineral Information Institute, 8307 Shaffer Pkwy., Littleton, CO 80127. Phone: 303-277-9190. Internet: http://www.mii.org.

National Hydrogen Association, 1211 Connecticut Ave., NW, Suite 600, Washington, DC 20036-2701. Phone: 202-223-5547. Internet: http://www.hydrogenassociation.org/.

Nickel Institute, Brookfield Place, 161 Bay St., Suite 2700, Toronto, ON, Canada M5J 2S1. Phone: 416-591-7999. Internet: http://www.nidi.org/.

Royal Society of Chemistry, Burlington House, Piccadilly, London, United Kingdom W1J 0BA. Phone: +44 (0)207 437 8656. Internet: http://www.rsc.org/.

Selenium-Tellurium Development Association, Cavite Economic Zone, Lot 6, Blk 1, Phase II Rosario, Cavite, Philippines. Phone: +63 46 437 2526. Internet: http://www.stda.org/.

Sulphur Institute, 1140 Connecticut Ave., NW, Suite 612, Washington, DC 20036. Phone: 202-331-9660. Internet: http://www.sulphurinstitute.org/.

Tantalum-Niobium International Study Center, Chaussée de Louvain 490, 1380 Lasne, Belgium. Phone: +32 2 649 51 58. Internet: http://tanb.org/.

U.S. Department of Energy (DOE), Lawrence Berkeley National Laboratory, 1 Cyclotron Rd., Berkeley, CA 94720. Phone: 510-486-4000. Internet: www.lbl.gov/.

U.S. Environmental Protection Agency (EPA), Ariel Rios Building, 1200 Pennsylvania Ave., NW, Washington, DC 20460. Phone: 202-272-0167. Internet: www.epa.gov/.

U.S. Geological Survey (USGS) National Center, 12201 Sunrise Valley Dr., Reston, VA 20192. Phone: 703-648-5953. Internet: www.usgs.gov/.

Index

Italic type indicates volume numbers; **boldface** indicates main entries. Illustrations are marked by (ill.)

E

F

H

H. *See* Hydrogen
Haber, Fritz, *2:* 395–396
Hafnium, *2:* **233–238,** 233 (ill.), 235 (ill.), 236 (ill.)
    discovery and naming of, *2:* 234
    isotopes of, *2:* 237
Hafnium-174, *2:* 237
Hafnium-176, *2:* 237
Hafnium-177, *2:* 237
Hafnium-178, *2:* 237
Hafnium-179, *2:* 237
Hafnium-180, *2:* 237
Hafnium boride, *2:* 238
Hafnium nitride, *2:* 238
Hafnium oxide, *2:* 238
Hahn, Otto, *3:* 475, 475 (ill.), 476–477
Haiti, elements in, *2:* 224
Half lives
    of actinium, *1:* 3
    of americium, *1:* 17
    of astatine, *1:* 42
    of berkelium, *1:* 51
    of californium, *1:* 97
    of curium, *1:* 162
    of einsteinium, *1:* 172, 173
    of fermium, *1:* 186
    of francium, *1:* 202
    of hafnium, *2:* 237
    of indium, *2:* 264
    of lanthanum, *2:* 304
    of neptunium, *2:* 371–372
    of plutonium, *3:* 441
    of promethium, *3:* 469
    of protactinium, *3:* 475
    of radioactive elements, *2:* 372
    of radon, *3:* 488, 489
    of rhenium, *3:* 494
    of selenium, *3:* 526
    of sodium, *3:* 548
    of tantalum, *3:* 571–572
    of technetium, *3:* 577, 578
Halite, *1:* 127, 128 (ill.); *3:* 547–548
Hall, Charles Martin, *1:* 7

Hall process, *1:* 9
Halogens. *See* Astatine; Bromine; Chlorine; Fluorine; Iodine
Hardness, *2:* 316
Hassium, *3:* 627, 630, 633
Hatchett, Charles, *2:* 383–384; *3:* 569–570
Hatmakers, *2:* 347
Haüy, René-Just, *1:* 54, 136
Hayyan, Abu Musa Jabir Ibn, *1:* 20
He. *See* Helium
Heart. *See* Cardiovascular system
Heat exchange, *3:* 456–457, 549–550
Helium, *2:* **239–246,** 239 (ill.)
    argon and, *1:* 27
    berkelium and, *1:* 49
    californium and, *1:* 95–96
    chemical properties of, *2:* 243, 245
    compounds of, *2:* 239, 243, 246, 363
    discovery and naming of, *2:* 239, 240–242, 293–294
    extraction of, *2:* 243–244
    health effects of, *2:* 246
    hydrogen and, *2:* 255
    hydrogen *vs.,* *2:* 258
    isotopes of, *2:* 243
    natural occurrence of, *2:* 243
    in periodic table, *2:* 239
    physical properties of, *2:* 239–240, 242
    uses of, *2:* 240, 243, 244–245, 244 (ill.), 246 (ill.)
Helium I, *2:* 242
Helium II, *2:* 242
Helium-3, *2:* 243
Helium-4, *2:* 243
Hematite. *See* Ferric oxide
Hemocyanin, *1:* 156, 156 (ill.)
Hemoglobin, *1:* 29, 155–156; *2:* 291
Heptachlor, *1:* 132
Herbicides. *See* Pesticides
Hevesy, George Charles de, *2:* 234
Hf. *See* Hafnium
Hg. *See* Mercury
HHS (Department of Health and Human Services). *See* U.S. Department of Health and Human Services

I

cerium and, *1*: 116

krypton and, *2*: 293, 296

krypton in, *2*: 297–299

lanthanum and, *2*: 304–305

lanthanum in, *2*: 301

magnesium and, *2*: 330

mercury in, *2*: 346, 346 (ill.)

neodymium in, *2*: 361

neon in, *1*: 26; *2*: 363, 367–368, 367 (ill.), 368 (ill.)

phosphorus and, *3*: 422, 423

praseodymium in, *3*: 464

promethium and, *3*: 471

radium and, *3*: 479

scandium in, *3*: 521

sodium in, *3*: 550

thorium in, *3*: 602 (ill.), 603 (ill.)

xenon and, *3*: 659 (ill.)

Lightbulbs, *2*: 394; *3*: 550, 646

Light-emitting diodes, *1*: 34, 184; *2*: 213–214, 213 (ill.), 265–266

See also Light and lighting

Lighters, *1*: 116 (ill.), 117; *2*: 301, 305

Lime. *See* Calcium oxide

Lime, slaked. *See* Calcium hydroxide

Limestone, *1*: 85–86, 88, 90–91, 91 (ill.), 105

alkaline earth metals and, *1*: 43

fluorspar and, *1*: 192

iron and, *2*: 288

potash and, *3*: 456

Limonite. *See* Hydrated ferric oxide

Linnaeite, *1*: 144

Lipowitz metal, *1*: 81

Liquid air

argon in, *1*: 25

helium in, *2*: 243–244

krypton in, *2*: 293, 297

neon in, *2*: 293

nitrogen in, *2*: 294, 393–394

oxygen in, *2*: 294, 411–412

xenon in, *2*: 293

Liquid crystal displays, *2*: 261, 263, 265

Lithium, *1*: 70, 116, 121; *2*: **315–320**, 315 (ill.), 372–373

discovery and naming of, *2*: 315–316

isotopes of, *2*: 317

uses and compounds of, *2*: 318–320

Lithium-6, *2*: 317

Lithium-7, *2*: 317

Lithium carbonate, *2*: 319, 320

Lithium chloride, *2*: 317

Lithium oxide, *2*: 317

Lithium stearate, *2*: 319–320

Lithium-ion batteries, *2*: 315, 318–319, 318 (ill.)

See also Batteries

Litvinenko, Alexander, *3*: 449

Liver, *2*: 229, 266

Löwig, Carl, *1*: 73, 74, 77

Lr. *See* Lawrencium

Lu. *See* Lutetium

Lubricants, *1*: 108; *2*: 355

Luminescence, *3*: 471, 479

Lungs. *See* Respiratory system

Lutectium. *See* Lutetium

Lutetium, *2*: **321–324**, 321 (ill.)

discovery and naming of, *2*: 321–323

isotopes of, *2*: 323–324

Lutetium-175, *2*: 323

Lutetium-176, *2*: 323, 324

Lutetium chloride, *2*: 324

Lutetium fluoride, *2*: 324

Lutetium oxide, *2*: 323

Lye. *See* Sodium hydroxide

M

Mackenzie, Kenneth R., *1*: 40

Macronutrients, *3*: 568

Magnesia. *See* Magnesium oxide

Magnesite. *See* Magnesium carbonate

Magnesium, *2*: **325–332**, 325 (ill.)

aluminum and, *1*: 10

argon and, *1*: 28

barium and, *1*: 43

beryllium and, *1*: 53, 55

bismuth and, *1*: 62

carbon and, *1*: 105

chemical properties of, *2*: 327–328

compounds of, *2*: 325, 326–328, 330–331

S

T

U

V

W